Inflation in
Engineering Economic Analysis

Inflation in Engineering Economic Analysis

BYRON W. JONES

Assistant Professor of Mechanical Engineering
Kansas State University

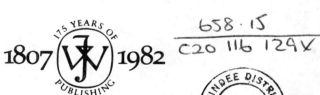

1807 1982

175 YEARS OF PUBLISHING

A Wiley-Interscience Publication
JOHN WILEY & SONS

New York   Chichester   Brisbane   Toronto   Singapore

Library of Congress Cataloging in Publication Data:

Jones, Byron W.
    Inflation in engineering economic analysis.

    "A Wiley-Interscience publication."
    Includes index.
    1. Engineering economy.   2. Inflation (Finance)
I. Title.
TA177.7.J66   658.1'5   81-13038
ISBN 0-471-09048-4   AACR2

Printed in the United States of America

10 9 8 7 6 5 4 3 2 1

*To
Melanie, Nickolas,
and Anthony*

Preface

This text is based on traditional engineering economic theory. However, some of the most basic concepts of this theory are examined to show how they are affected by inflation and how also the resulting engineering economic calculations and decisions are affected. This examination starts with the practice of using the dollar (or other currency) as the unit of measure for economic value. The dollar is replaced by an artificial "constant dollar" that establishes value in terms of goods and services. With this background, I review the principle of the time value of money, the central concept of basic engineering economics. The time value of money is usually explained as the cost for investment capital in terms of either a money market price or the return possible for alternative investments. It is argued here that the traditional form of the time value of money is as much, if not more, a correction for inflation and the resulting decline in the value of the dollar as it is a measure of the real time cost of capital. Just as most engineering economics texts call for risks to be evaluated separately and not be combined with the time value of money, it is suggested that inflation be evaluated separately and not combined with the real cost of capital. Along with this presentation, I also show that correct analyses can be made using traditional definitions of the time value of money and dollar measures of economic value if care is used to consistently estimate cash flows, the time value of money, and inflation.

I have taught an introductory course in engineering economics for several years and have been amazed at the lack of discussion of inflation in the texts available for such a course. Many texts do not even mention inflation, and those that do so usually have a very limited discussion and seldom adequately portray just how important inflation is in both theory and application of engineering economics. This book originated as a set of notes to provide what I felt was necessary supplementary material for an introductory course. Those notes have been expanded to be of use for a wider range of applications. However, this book is intended not as a comprehensive treatise on inflation or engineering economics, but rather to explain how basic engineering economics calculations can be made to reflect the reality of inflation in a theoretically sound yet reasonably manageable manner.

It is anticipated that there will be several applications for this text. It can be used in an introductory course in engineering economics to supplement one of the many fine texts already available. It can be used as a text for a separate course on inflation in engineering economics following an introductory course that does not deal with inflation in any detail. It can also be of use to practicing engineers who are familiar with engineering economics but who would like to be able to account for inflation more properly in their economic analyses. Throughout the book it is assumed that the user is familiar with basic engineering economic principles, particularly the time value of money, and with the present worth, annual cash flow, and internal rate of return methods of making economic evaluations. These principles and calculations are reviewed, not for the purpose of educating the uninitiated, but rather for the purpose of explaining how they are affected by inflation.

Chapter 1 gives a qualitative description of inflation, distinguishing between cost inflation and price inflation. Chapter 2 describes inflation in more precise mathematical terms, thus allowing a quantitative evaluation of inflation. The concept of a constant dollar is presented here for use throughout the rest of the text. Chapter 3 is devoted to the important concept of the time value of money. Inflation is shown to be an important component of the time value of money, and the important difference between the cost of money and the real cost of capital is explained. Chapter 4 discusses how engineering economic evaluations may be made by using the real cost of capital and constant dollar expression of cash flows or the cost of money and dollar expression of cash flows. The importance of properly and consistently accounting for inflation when setting up solutions is emphasized. Chapters 5 through 7 explain how the concepts developed in Chapters 1 through 4 apply to the three common methods for making engineering economic comparisons: present worth, annual cash flow, and the internal rate of return. Chapter 8 describes how inflation affects taxes. United States federal tax codes are used for this presentation. Chapter 9 shows how inflation can be considered a factor in an uncertainty analysis and also how to properly account for inflation in sensitivity analyses involving price changes. Chapter 10 briefly reviews the difficulty involved in making economic calculations when inflation is severe. Elsewhere in the text inflation is assumed to be significant but not too great. The theory for using inflation rates and costs of money or capital that vary with time is developed in Chapter 11. This theory allows inflation rates to be projected as varying in the future rather than remaining constant during the entire period of an economic analysis. Chapter 12 presents an alternative to describing inflation in discrete time form.

Inflation relationships analogous to continuously compounded interest are derived.

Example problems are presented at the end of each chapter. This is to avoid confusion between general theoretical concepts and specific applications. Instructors using this text may wish to incorporate the examples into the theoretical developments for classroom presentations. This separation allows instructors to tailor the use of examples to fit their own teaching methods and their own balance of theory and application.

It is my opinion that Chapters 1 through 7 represent the minimum exposure engineers should have in dealing with inflation in engineering economic analysis. Chapters 8 and 9 pertain to material that is included in most introductory courses and probably should also be covered by students in these courses.

Chapters 1 through 9 are intended to be covered sequentially. Chapters 10 and 11 might be considered optional material and probably do not need to be covered unless the text is being used for a separate course on inflation. Chapter 11 should be of particular benefit for those students who intend to continue on to more advanced studies in engineering economics. Chapter 12 is included and mainly for those who expect to routinely make economic analyses and would like to simplify some of the calculations. Finally, attention is called to Appendix A, which answers the inevitable question of what to do if we have deflation rather than inflation.

No book is the result of the efforts of only one person. I would like to take this opportunity to thank all those people who have contributed in one way or another. I hesitate to name individuals because it is impossible not to leave someone out. I would particularly like to thank those students who were persistent in their questioning of why inflation appeared to be ignored and who consequently served as a catalyst for the beginning of this text, and I thank those students who served as test subjects for much of this material. Special thanks go to my family, who have been continually forgiving of my chronic underestimation of the amount of my time this task would take from them.

BYRON W. JONES

Manhattan, Kansas
November 1981

Contents

Inflation in
Engineering Economic Analysis

CHAPTER ONE

Introduction

1.1 MONEY AND VALUE

Understanding inflation requires an understanding of money. The U.S. dollar* has no significant intrinsic value. That is, the materials in a dollar have very little value, no more than the paper they are printed on. The coins that we use have intrinsic values that are generally much less than their face value. Most of the money in the United States, checking accounts, exists only as entries in account books. Yet a dollar undeniably does have value. Dollars can be exchanged for valuable goods and services.

What is the dollar, then, and what makes it valuable? A complete answer to this question is probably impossible. Any attempt to do so would entail the writing of another whole book. To greatly simplify the answer, dollars can be considered as promisory notes of the U.S. government and its citizens to provide goods and services in return for these notes. Dollars have value because the law gives them legal tender status. Dollars may be used in payment of all public and private debts. However, both of these factors give the dollar value in terms of dollars. A $1 promisory note entitles the holder to goods and services equal in value to $1. This does little to help us understand why the dollar has value.

When all is said and done, the value of the dollar is based largely on faith. We accept dollars in exchange for goods and services because we believe we can exchange the dollars again for other goods and services. As long as this faith exists, money has a value. The U.S. money system is what is called a *managed money system*. The government, through the banking system, takes an active role in trying to maintain a stable value of the dollar. But even this process is not well understood, and the value of the dollar does change. There is nothing that sets the actual value of a dollar. It can and will change in value.

*The dollar is used throughout this text. However, the theory and discussions presented in the text apply equally to most modern currencies.

1

The dollar appears to have a very tenuous existence. It has no intrinsic value, the value it does have is based largely on faith, and its value does change. In spite of these limitations, the dollar is our primary standard of value. Just as the meter is used to measure distance, the kilogram is used to measure mass, and other units are used for other types of measurement, the dollar is used to measure value. Fortunately, the standards for physical units do not vary like the dollar does.

It may be difficult to fully appreciate the importance of the dollar as a standard of value. Imagine the difficulties that would arise if no such standard existed. The value of every good and service would need to be expressed in terms of every other good and service for trades to be made. With the dollar, the value of all goods and services can be expressed in terms of a single unit. The values of a house, a gold ring, a bag of rice, an automobile, and so on may all be expressed in terms of this single, readily recognized unit of measure.

The dollar is more than just a measure of value, however. It *is* value. It can be exchanged for goods and services. It provides a means of transporting value and dividing value into appropriate quantities. This function of money eliminates the necessity of every transaction being a trade of goods and services; a wheat farmer who needs a plow does not have to find an implement dealer who needs grain, and a carpenter who needs a tooth extracted does not have to find a dentist who needs a house built. An economy without some form of money to expedite such transactions would certainly be inefficient. To be sure, societies have existed that relied on barter. However, these have been primitive societies or societies in which the monetary system was destroyed by war or other disruptions.

1.2 MONEY IN ENGINEERING ECONOMIC ANALYSIS

Just as money serves as a standard of value for trade, it serves the same purpose in engineering economic analysis. A single investment may consist of dozens or even hundreds of diverse goods and services. The return from the investment may consist of different, equally diverse goods and services. For example, building even a small manufacturing plant requires wood, structural steel, pipe, valves, wiring, switches, concrete, shingles, lights, heaters, glass, and so on, in addition to various skilled laborers. Machines, electricity, raw materials, and skilled labor are needed to operate the plant. The plant may produce many different products or, perhaps, only one product. In any case, it would be virtually impossible to determine whether such an investment were sound unless all the inputs and outputs could be expressed in terms of a single common unit. The same is true for almost every economic analysis. No meaningful economic judgments can be made until everything is expressed in dollars.

There is a danger in using dollars for economic analysis, however. We tend to get used to thinking of economic analysis in terms of dollars. An investment costs so many dollars and returns so many dollars. It is easy to forget that the cost of an investment is the goods and services that go into it and the return on an investment is the goods and services it provides. The dollars are not the cost or the return but only measures of these quantities. Overlooking this simple, seemingly obvious point causes untold numbers of people and companies to unwisely allocate their money and measure their economic success.

As long as the dollar remains a consistent and valid measure of value, it is reasonable to make economic analyses strictly in dollar terms. When the amount of the value represented by a dollar is not constant, an economic analysis made in dollar terms is likely to be incorrect. Inflation results in the dollar being an inconsistent measure of value. The purpose of this book is to show how this deficiency can be corrected so that valid economic analysis may be made.

1.3   INFLATION

Inflation is defined here as an increase in the prices paid for goods and services. The increase in prices can result from either cost inflation or price inflation, or combinations of the two.

Cost inflation is the result of real cost increases to produce goods and services; that is, more inputs in terms of labor and capital are needed to produce a given amount of goods and services. Society as a whole must work harder and invest more to produce the same goods and services as before. Cost inflation results from a number of causes. Depletion of natural resources can result in cost inflation. The easily discovered, cheaply produced resources are used first. When these are used up, more expensive resources must be utilized, resulting in a real cost increase to society. Cost inflation can result from natural disasters and catastrophes. Drought, floods, blizzards, and so on all make it more difficult and costly for an economy to function and consequently increase the cost of producing goods and services. Social and political disruptions have a similar effect on the economy. In fact, anything that has a negative effect on productivity tends to cause real cost increases and cost inflation. Conversely, anything that increases productivity results in real cost decreases to society as a whole and tends to cause lower real

prices. Throughout much of human history, social and technological
advances have resulted in real cost decreases. However, this past experience
does not mean that real cost increases are not occurring today.

Price inflation is something quite different. Since the value of the dollar is
largely an illusionary quantity and there is nothing to specifically set this
value, this value is fairly free to float. There is no limitation that a dollar must
represent any given quantity of goods and services or that it represents any
given amount of labor and investment to produce goods and services.
Consequently, the prices of all goods and services may rise or fall even when
their real costs do not change. Such behavior in the economy is called *price
inflation*. The causes and cures of price inflation are not perfectly understood
but are the topic of endless debate and discussion. There are almost as many
theories of price inflation as there are people discussing it. It is beyond the
scope of this text to discuss all these theories in detail. However, it may be
useful to review some of the basic concepts.

The most basic money equation in the economy takes the form

$$p \times N = M \times V \tag{1.1}$$

where p is the price level in dollars/unit of value, N is the rate at which
transactions take place in units of value/year, M is the amount of money held
in the economy in dollars and V is the velocity of money (i.e., the rate at
which money is spent for a given amount of money held) in units of
(dollars/year)/dollars. One theory of price inflation holds that inflation is due
mainly to increases in the amount of money M held in the economy. If this
so-called supply of money is increased, prices are forced to increase also.
The additional money may come from either government issues or by
individuals through banks. The supply of money in the United States is
regulated to a certain extent by the government through the federal reserve
system. However, this control is by no means absolute. A second theory of
price inflation holds that inflation is due largely to changes in V. People and
businesses may change their spending habits and spend their money more
readily and hold less dollar reserves. Such an increase in the money velocity
is claimed to force the prices upward. Other theories of price inflation hold
that neither M nor V are controlling factors but rather respond to other
forces. These forces may be the inherent structure in our economy that resists
wage and price decreases but readily allows increases. These theories point
out that Equation 1.1 is not necessarily a causal relationship but is simply a
"truism" that must be satisfied much the same as an energy balance or mass
balance must be satisfied in physical systems.

Price inflation and cost inflation are very different. Cost inflation is not
directly related to money. It could take place, at least in theory, in economies

that do not use money. An important point often overlooked is that cost inflation must be absorbed somewhere in the economy. It is the result of real cost increases, and somebody (if not everybody) must either work more to produce the same goods and services or earn fewer goods and services for the same work. Price inflation, on the other hand, does not entail real cost increases. It is strictly a money phenomenon and it is not necessary that any increased costs be absorbed. People pay more for everything but also earn more for their work. In reality, however, price inflation inevitably impacts different people differently. Many wages and prices are fixed in dollar terms rather than value terms and changing prices changes the value of the associated transactions. Also, it should be remembered that prices are continually changing in our economy and are seldom likely to change by the same amounts for all items simultaneously.

Cost inflation and price inflation may occur at the same time and may have an interactive relationship that tends to accelerate inflation. Cost inflation may occur, resulting in decreased effective earnings for some people or companies. If they have sufficient market power, they may pass these costs on to others to maintain their own real income at previous levels. If other people also have sufficient market power, they may also pass the costs on. Sooner or later the costs must be asorbed; but if everybody is able to keep passing the increases on, a wage-price spiral can develop. The initial price increases may have been due to cost inflation, but the wage price spiral that results if nobody is willing to absorb any of the real cost increase is a price inflation phenomenon.

Fortunately, the engineering economist seldom needs to be concerned with the causes of inflation and all the different theories. It is important that the engineering economist be concerned with the effect of inflation on economic analysis. The effect of either price or cost inflation is to seriously limit the usefulness of the dollar as a measure of value. Inflation results in the dollar having a different value at different times. Economic analyses made strictly on a dollar basis do not account for this fact. It is not fair to compare a dollar today to a dollar in the past; they do not represent the same amount of value. Only a fool would compare two lengths, one measured in meters and one measured in feet, or two weights, one measured in pounds and one in newtons. Both quantities must be expressed in the same unit before a meaningful comparison can be made. However, dollars at different points separated by time periods of significant inflation are just as different as meters and feet. Economic comparisons involving these two different measures of value can be just as misleading as comparing quantities measured in feet to quantities measured in meters. Unfortunately, most engineering economic analysis involves transactions that occur over a few years to many years in time. Dollars are a very questionable measure of value for such analysis.

If dollars are not valid measures of value, then what is? There is no easy answer to this question. There is no fixed standard of value that can be used for all measures of value. The engineering economist must recognize that inflation changes the value of the dollar and correct for these differences if accurate analyses are to be made. The corrections are not always perfect, but they are much better than ignoring inflation and its effect on our value standard. The purposes of the remaining chapters herein are to show how these corrections can be made, to show the distortions that can be caused by inflation, and to show the limitations of economic analysis when significant inflation exists.

1.4 EXAMPLES

The existence of both cost and price inflation can be readily seen. Cost inflation is apparent in a number of natural resources where the cheap, high quality sources have been used up. Fortunately, most of the cost inflation due to depletion has been offset by technological advances. Price inflation, too, is not difficult to spot. This inflation is readily seen in the tremendous rise in prices over long periods of time that have at the same time resulted in improved living standards. The following examples point out particular instances where cost and price inflation can be seen.

Example 1.1

The effects of depletion of the crude oil resources in the United States are easily seen. The depth at which exploration must be carried out has steadily increased, as is shown in Figure 1.1. Drilling to greater depths unquestionably results in greater costs. In addition to being at greater depths, there is a tendency for new fields to be of lower value. Figure 1.2 shows that the fraction of the new fields discovered that turn out to be commercially profitable has steadily declined. This decline also increases the cost of oil.

The fact that these two exploration cost factors have been increasing does not necessarily mean that the total cost of producing oil increased over this time period. Many other factors are also important, and the net cost may not have increased as a result of these two factors alone. However, they are one element of the cost of producing goods and services for our society, and they have contributed to cost inflation.

Example 1.2

The effects of depletion can be seen with many other natural resources. Copper ore is a typical example. The average grade of copper ore in

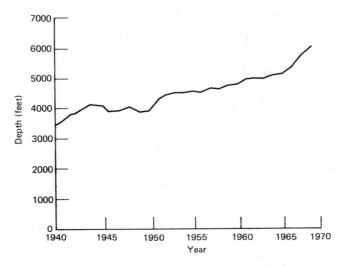

Figure 1.1   Average depth of exploratory wells.[1]

producing mines in the United States is presently about 7/10%. The recoverable content of this ore is about 5/10%.[2] Some mines utilize ore at grades significantly less than this amount. In 1900 copper ore below a grade of 3% was not normally mined.[3] Figure 1.3 shows how the amount of ore that must be handled varies with the ore grade. Figure 1.4 shows how the energy

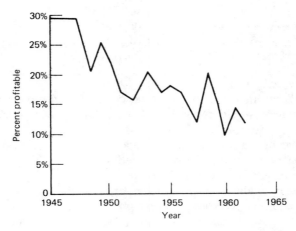

Figure 1.2   Fraction of the new fields discovered that were commercially profitable[1] (ultimate size greater than 25 million barrels).

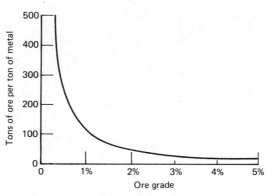

Figure 1.3 Amout of ore that must be handled to produce 1 ton of metal (assuming that the last 1/10% is unrecoverable).

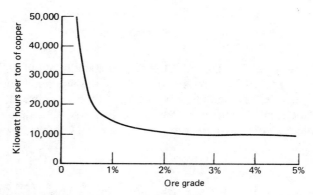

Figure 1.4 Energy requirements to produce copper (based on Figure 1.3 and the estimate of 48 kW · h/ton of ore plus 8715 kW · h/ton of copper[4]).

requirements, an important part of total cost, vary with ore concentration. Copper producers would certainly rather produce 3% ore than they would 7/10% ore.

The decreasing grade of available copper ore clearly tends to increase the cost of copper and contributes to cost inflation. However, the copper industry is a good example of how cost inflation pressures due to depletion have been largely offset by technological advances. It is claimed that the present-day cost of extracting copper from ore with a grade of 5/10% is no higher than the cost to extract copper from ores with grades of 5–7% a hundred years ago when this cost is measured in goods and services.[5]

TABLE 1.1 Typical Prices at the Turn of the Century[6]

Item	Price
Coffee	17–35¢/lb
Raisins	6–11¢/lb
Oatmeal	3¢/lb
Broom	25¢
Shovel	80¢
Pocket knife	15–50¢
Wood stove	$3–$15
Man's suit (wool)	$5–$6
Quilt	50¢–$1.75
Drafter's instrument set	$3.50
Shotgun	$5–$10
Rocking chair (solid wood)	$2
Roll top desk	$15–$20

Example 1.3

In addition to examples of cost inflation, it is easy to see that significant price inflation has occurred over the years. Table 1.1 lists the typical prices of some items around the turn of the century. These items are also common today. Comparing the prices in Table 1.1 to current prices, it is seen that these items have increased in price by factors of approximately 10–100. These items are a small part of the total cost of living, today and then. However, they are indicative of how prices have changed during this century. It does not cost 10–100 times more for these items today when the cost is measured in terms the human effort required to earn them. This increase in price certainly does not indicate a comparable decrease in the standard of living, but rather long-term price inflation.

REFERENCES

1. *Petroleum Facts and Figures*, 1971 Ed., American Petroleum Institute, New York, 1971.
2. *Copper: The Next Fifteen Years*, W. Gluschke, J. Shaw, and B. Varon, Reidel, Publ., Dordrecht, Boston, London, 1979.
3. "Limits to Exploitation of Nonrenewable Resources," E. Cook, *Science*, **191**, 677–682 (1976).

4. "Ore Grade, Metal Production, and Energy," N. Page, S. Creasey, *Journal of Research, U.S. Geological Survey*, Vol. 3, No. 1, 9–13, 1975.

5. *The Economics of Natural Resources*, F. Banks, Plenun, New York, 1976.

6. *1897 Sears Roebuck Catalogue*, F. L. Israel, ed., Chelsea House, New York, 1968.

CHAPTER TWO

Quantifying Inflation

Inflation was defined in Chapter 1 as an increase in the price of goods and services. This simple definition does little to help us include inflation in our engineering economic analyses or to overcome the critical problem of the dollar as a unit of measure that has a changing value. In this chapter a mathematical definition of inflation is given and means for quantifying these expressions presented.

2.1 CONSTANT VALUE DOLLAR

From the point of view of economic analysis, the ideal dollar would always represent a given amount of value. No such unit of money has ever existed or is likely ever to exist. However, the dollar does represent some amount of value at a given point in time. This value of the dollar is represented by $\bar{p}(t)$, the average price of goods and services at time t.* The inverse of $\bar{p}(t)$ may be regarded as the buying power of a dollar. Inflation is then an increase in the prices of goods and services $\bar{p}(t)$. This definition is the same as saying inflation is a decrease in the buying power of the dollar.

It is more convenient to state the definition of inflation in terms of the rate at which prices change. Mathematically, inflation is defined by

$$\bar{p}(t + \Delta t) = \bar{p}(t) \cdot [1 + f] \qquad (2.1)$$

where f is the inflation rate. The inflation rate is seen to be the fractional change in prices for a time period of length Δt. If the inflation rate is assumed to stay constant (an important assumption), then Equation 2.1 may be extended to any number of time periods Δt:

---

*See Appendix 2.2 for further discussion of this definition.

$$\bar{p}(t + n \cdot \Delta t) = \bar{p}(t) \cdot [1 + f]^n \qquad (2.2)$$

where n is the number of time periods.

Valid economic analysis requires a consistent measure of value, but the buying power of the dollar varies because of inflation. Somehow adjustments must be made in dollar cash flows to account for this change in value. The cash flow required to buy goods and services of value V is given by

$$Y^d(t) = V \cdot \bar{p}(t) \qquad (2.3)$$

where $Y^d(t)$ is the cash flow in dollars.

The ratio

$$\frac{Y^d(t)}{\bar{p}(t)}$$

then stays constant for a constant amount of value, V. This ratio can then be used to determine the cash flows required to purchase the same value of goods and services at two different points in time since

$$\frac{Y^d(t_1)}{\bar{p}(t_1)} = \frac{Y^d(t_2)}{\bar{p}(t_2)} \qquad (2.4)$$

Equation 2.4 allows for comparison of all cash flows on a common basis. Let t_0 be some arbitrary reference time. Then any cash flow at any point in time may be expressed as the cash flow that would have resulted with the average prices existing at t_0. This cash flow, called the *constant dollar cash flow*, can be calculated from Equation 2.4:

$$Y^c(t) = Y^d(t) \cdot \frac{\bar{p}(t_0)}{\bar{p}(t)} \qquad (2.5)$$

where Y^c is the constant dollar cash flow. Now, a number of cash flows of dollars of different values may all be expressed in a constant dollar of the same value that is the consistent measure of value being sought.

The constant dollar calculation in Equation 2.5 requires only that the ratio

$$\frac{\bar{p}(t_0)}{\bar{p}(t)}$$

be known; values for the prices need not be known. This ratio can be calculated from Equation 2.2:

$$\frac{\bar{p}(t_0)}{\bar{p}(t)} = \frac{1}{[1+f]^n} \tag{2.6}$$

where

$$n = \frac{t - t_0}{\Delta t} \tag{2.7}$$

The mathematical form of Equation 2.6 is identical to the present worth factor commonly used in engineering economics. This expression and the present worth factor are analogous in some ways but should not be confused with each other.

This ratio may be regarded as a "deflator" converting dollars of one value to dollars of a different value much the same as a conversion factor is used to convert feet to meters or gallons to liters. Mathematically, the deflator is defined as

$$D(t) = \frac{\bar{p}(t_0)}{\bar{p}(t)} \tag{2.8}$$

where $D(t)$ is the deflator and has units of dollar of constant value per dollar at time t. To keep the two different types of dollar straight, the symbol \$ is used here to represent an actual dollar cash flow regardless of the value of that dollar, and the symbol $\bar{\$}$ is used to express the corresponding cash flow expressed in a reference year's dollar. The deflator then has units of $\bar{\$}/\$$. With the use of the deflator, Equation 2.6 becomes

$$D(t) = \frac{1}{[1+f]^n} \tag{2.9}$$

when f is constant, and Equation 2.5 becomes

$$Y^c(t) = D(t) \cdot Y^d(t) \tag{2.10}$$

Equation 2.10 is valid even if the inflation rate is not constant, but $D(t)$ must be found from Equation 2.8 and not 2.9. The inflation rate is assumed constant for most of the material that follows; however, Chapter 11 is devoted to the real problem of varying inflation rates.

2.2 PRICE INDEXES

For the preceding discussion to be more than an academic exercise, some further quantification of inflation is in order. Determinination of an inflation rate or expressing a cash flow in constant dollar terms requires estimation of the ratio

$$\frac{\bar{p}(t_0)}{\bar{p}(t)}$$

It is difficult to directly estimate this ratio since there is no other common measure of value except the dollar and it is impossible to use a standard to measure changes in that standard. The data that are probably the most useful for determining this ratio are the various price indexes.

A price index is a simple concept and is defined as

$$I(t) = \frac{\bar{p}'(t)}{\bar{p}'(t_b)} \tag{2.11}$$

where $I(t)$ is the price index, t_b is a "base" year and the ' notation indicates that $\bar{p}$ is for a set of goods and services and not necessarily for all goods and services. The price index thus compares the cost of goods and services at time t to the price of the same goods and services at a base year t_b. Price indexes may be tabulated for different groupings of goods and services, ranging from groups including all goods and services to groups of particular items (e.g., electrical equipment). Also, the base year t_b should not be confused with the reference year t_0 used in the constant dollar calculation.

The more comprehensive price indexes should give a reasonable estimate of the change in the overall buying power of the dollar. Assuming that

$$\frac{\bar{p}(t_0)}{\bar{p}(t)} = \frac{\bar{p}'(t_0)}{\bar{p}'(t)} \tag{2.12}$$

it follows that

$$\frac{\bar{p}(t_0)}{\bar{p}(t)} = \frac{I(t_0) \cdot \bar{p}'(t_b)}{I(t) \cdot \bar{p}'(t_b)}$$

or

$$\frac{\bar{p}(t_0)}{\bar{p}(t)} = \frac{I(t_0)}{I(t)} \qquad\qquad (2.13)$$

Compilation of price indexes is no simple task. The theory and calculations involved are outside the scope of this text and generally outside the scope of engineering tasks. However, users of price indexes should be aware of the difficulties in developing accurate price indexes so as to realize some of their limitations. Some of the more important difficulties are:

1. It is not always possible to determine exactly the prices of goods and services.
2. At any given time a given good or service sells at different prices at different locations. Some weighting scheme is required to average these prices.
3. The prices of different goods and services used to calculate a price index cannot be averaged directly to determine a price index. Some means of weighting the price change for each item in the group must be used.
4. The quality of particular goods or services varies. It is difficult to separate price changes due to changes in quality from other price changes.
5. New products come on the market for which there is no base to which prices may be compared.

These difficulties limit the accuracy with which price indexes can quantify inflation. However, this quantification is certainly no less accurate than many others that are frequently encountered in economic analysis (e.g., the value of employee safety).

The U.S. government tabulates two major categories of price indexes on a routine basis. These are the Wholesale Price Indexes (WPIs) and the Consumer Price Indexes (CPIs). Each category covers specific products and has its areas of usefulness.

The WPIs measure prices in the primary markets compared with prices prevailing for comparable commodities in a given base period.[1] This base time period is updated periodically. The prices used in developing the indexes are those for large quantity purchases and are not necessarily the prices at the local wholesale level. As far as possible, the prices represent the first transaction after manufacture. The indexes are designed to reflect discounts, rebates, and so on that are involved in the transaction. Also, the indexes are designed to account for changes in quality to the fullest extent possible. The WPIs are developed at a very disaggregate level and then combined to obtain broader based averages. The price data are combined by using weights based on value of shipments. Table D.5 (in Appendix D) presents the commodity grouping in the index. Table D.6 shows historic values of these price indexes. Aggregate values are presented in Table 2.1.

The CPIs measure changes in prices of goods and services individuals and families buy to live (e.g., food, clothing, housing, medicine, recreation, transportation).[2] The prices include all sales, excise, and other taxes directly associated with the purchase. The indexes measure only prices and do not account for changes in quality. Prices are obtained by sampling stores and service establishments in a number of cities. The CPIs are also calculated at a fairly disaggregate level and then combined to obtain averages. Table D.2 shows the historic values of these indexes. Aggregate values are given in Table 2.1.

A third measure of prices is the Gross National Product Implicit Price Deflator (GNP-IPD). The GNP-IPD is based on the entire Gross National Product (GNP) of the United States and is designed to reflect what this group of goods and services would have cost to purchase at prices prevailing in the base year.[3] The cost of the components of the GNP are evaluated at today's prices and at the base year's prices at as disaggregate a level as the available price indexes permit. The ratio of the total cost for today and for the base year is the GNP-IPD. Since it includes all goods and services produced, it is the widest based of the indexes. Historic values for the GNP-IPD are presented in Table 2.1 and Table D.1.

2.3   USING PRICE INDEXES

Price indexes may be used to measure inflation, to convert dollar cash flows to constant dollar cash flows, and to show how the price of one item changes relative to others. All these calculations are similar but have different purposes.

Use of a price index to measure inflation requires that the index reflect the overall change in the buying power of the dollar. It is seen in Table 2.1 that, even at the most aggregate levels, the WPI, the CPI, and the GNP-IPD do not show the same changes. Which one of these is the best measure of inflation is an open question. It is the author's opinion that the GNP-IPD is the preferred measure since it is the most broadly based.

Regardless of the index chosen, the inflation rate is determined by first using the index to measure the ratio of prices at two different times as in Equation 2.13 and then using Equation 2.1 to solve for the inflation rate:

$$f = \frac{I(t)}{I(t_0)} - 1 \qquad (2.14)$$

TABLE 2.1 Price Indexes[a]

Year	Consumer Price Index	Wholesale Price Index[b]	Gross National Product Implicit Price Deflator
1947	66.9	76.5	62.9
1948	72.1	82.8	67.2
1949	71.4	78.7	66.6
1950	72.1	81.8	67.8
1951	77.8	91.1	72.5
1952	79.5	88.6	73.4
1953	80.1	87.4	74.6
1954	80.5	87.6	75.6
1955	80.2	87.8	77.2
1956	81.4	90.7	79.6
1957	84.3	93.3	82.3
1958	86.6	94.6	83.7
1959	87.7	94.8	85.4
1960	88.7	94.9	87.0
1961	89.6	94.5	87.7
1962	90.6	94.8	89.4
1963	91.7	94.5	90.6
1964	92.9	94.7	92.0
1965	94.5	96.6	94.1
1966	97.2	99.8	97.2
1967	100.0	100.0	100.0
1968	104.2	102.5	104.6
1969	109.8	106.5	109.7
1970	116.3	110.4	115.7
1971	121.3	113.9	121.5
1972	125.3	119.1	126.6
1973	133.1	134.7	133.9
1974	147.7	160.1	146.8
1975	161.2	174.9	161.0
1976	170.5	183.0	169.2
1977	181.5	194.2	179.4
1978	195.4	209.3	192.4
1979	217.4	235.4	209.6
1980	246.8	268.8	224.6

[a] Averages for year.

[b] Producer Price Index beginning 1978.

Source: *Business Conditions Digest*, U.S. Department of Commerce, Bureau of Economic Analysis.

17

and

$$\Delta t = t - t_0 \qquad (2.15)$$

It is customary to state inflation rates on an annual basis ($\Delta t = 1$ yr). Other time periods may also be used, as is described in Appendix 2.1.

Constant dollar cash flow expressions require that all dollar cash flows be expressed in the corresponding cash flow for a given reference year. Consequently, a single reference year t_0 must be specified for these calculation. This reference year is arbitrary and should be selected to be as convenient as possible. Today, meaning the time at which an analysis is made, is frequently selected as the reference time. The price indexes may be used to estimate the ratio

$$\frac{\bar{p}(t_0)}{\bar{p}(t)}$$

as in Equation 2.13 and then applied to Equation 2.5 to make the constant dollar conversion. It is often more convenient to use the preceding ratio to calculate the deflator (Equation 2.8) and then use it to make the constant dollar conversion in Equation 2.10. When using price indexes to convert cash flows, an aggregate index should be used since the conversion represents the overall change in the buying power of a dollar due to inflation.

It should be evident from Tables D.3 and D.6 that all prices do not change by proportionate amounts. This situation generally prevails regardless of whether there is inflation. However, individual price changes are difficult to assess when the value of the dollar changes also. That is, it is difficult to separate price changes due to changes in the real cost of a product from price changes due to inflation. One approach to this problem is to convert prices to constant dollar terms:

$$p_j^c(t) = p_j^d(t) \cdot \frac{\bar{p}(t_0)}{\bar{p}(t)} \qquad (2.16)$$

where p_j^c is the price of item j in terms of the reference year dollar and $p_j^d(t)$ is the dollar price of j at time t. The ratio

$$\frac{\bar{p}(t_0)}{\bar{p}(t)}$$

is determined in the same manner as has been done before. A change in p^c is referred to as a "real" price change as it represents a change in the value cost of item j.

Equation 2.16 is very useful when j represents a single item for which price data are available. It becomes less useful when a whole category of items is involved (e.g., pipe fittings). A single price does not apply in this situation. Instead, a price index can be used to measure the average price of items in the category. The ratio

$$\frac{I_j(t)}{I_j(t_0)}$$

where I_j is the price index for category j; describes how the dollar price of the items has changed from year t_0 to year t. This ratio can then be multiplied by the ratio

$$\frac{\bar{p}(t_0)}{\bar{p}(t)}$$

which describes how the value of the dollar has changed during the same time period. The result is the relative price change for category j:

$$p_j^r\left(\frac{t}{t_0}\right) = \frac{I_j(t)}{I_j(t_0)} \cdot \frac{\bar{p}(t_0)}{\bar{p}(t)} \tag{2.17}$$

where p_j^r is analogous to p_j^r in Equation 2.16 except that it is dimensionless and does not necessarily represent a single item. It will be 1.0 when there is no real price change between t_0 and t. It is the ratio of price changes in items j to the overall price changes in the economy. It should also be evident that the ratio

$$\frac{\bar{p}(t_0)}{\bar{p}(t)}$$

may be replaced with $D(t)$ in either of these equations.

2.4 DEALING WITH THE FUTURE

The discussion thus far in this chapter has been oriented toward dealing with the past. Engineering economics deals with the future. The past is important in that it determines the present and may provide information from which we can learn important lessons. The important considerations with respect to inflation are how it affects our future actions. Nearly all engineering economics problems require predicting such future events as the cost of a project, the demand for a given product, or how much throughput to expect. Many problems contain important variables for which substantial uncertainty exists. Inflation is no different. No one can know the exact future of inflation rates. However, this uncertainty does not mean that inflation is not an important variable in many analyses. Methods and procedures for predicting future inflation rates are not dealt with in this text. Discussions are limited to dealing with inflation once estimates of inflation rates are made and, in the case of Chapter 9, dealing with uncertainty in the predicted inflation rates.

The relationships and theory presented thus far in this chapter are just as valid for the future as they are for the past except that future inflation rates are based on forecasts whereas past inflation rates are based, hopefully, on sound data. For future purposes, it is usually sufficient to use a single inflation rate and assume that it stays constant. Use of different inflation rates for different periods of time is perfectly valid, but the added complexity may not be justified in view of the uncertainty in future inflation. The majority of the discussion in following chapters assumes a constant inflation rate, at least for a specified time period. In any event, it should be realized exactly what an inflation rate forecast is. It is an estimate of how the value represented by a dollar will change with time. Along with inflation forecasts, it is important to estimate the future prices of specific items included in the analysis. These individual prices do not necessarily change at a rate equal to the inflation rate, and serious errors can result if these differences are not recognized.

2.5   EXAMPLES

Example 2.1

During a period of time when the inflation rate is 10% each year, how much will the value of the dollar decline in five years?

Let us arbitrarily define the price of a unit of value at time zero as $1.0/unit. The price of the same unit of value after five years is then given by

$$\frac{\bar{p}(5)}{\bar{p}(0)} = [1.0 + f]^n$$

and

$$n = \frac{5 \text{ yr} - 0 \text{ yr}}{1 \text{ yr}} = 5$$

giving

$$\bar{p}(5) = [\$1.0/\text{unit}] \times [1.1]^5 = \$1.61/\text{unit}$$

Thus it requires 61% more dollars to buy the same unit of value after five years. Another way of stating the same result is to say that one dollar will buy one unit of value at time zero and only

$$\frac{1}{1.61} = 0.62$$

units of value after five years. The value of the dollar declined 38% during this time period.

Note that the inflation rate was expressed as an annual percentage in this problem. This form is common. It is generally safe to assume that a stated inflation rate is for one year unless another time period is specified. Also, for the inflation rate to be used in the equations, it must be expressed as a fraction and not a percentage.

Example 2.2

A year's supply of feedstock for a chemical plant has been purchased on contract. The contract calls for payments at the end of each month as follows:

January—$750,000	July—$800,000
February—775,000	August—810,000
March—795,000	September—810,000
April—800,000	October—800,000
May—800,000	November—780,000
June—800,000	December—760,000

The inflation rate is also expected to be $1\frac{1}{2}\%$ each month during this year. What are the constant dollar cash flows for the preceding dollar cash flows? Use January 1 as the reference time.

The constant dollar cash flow can be calculated by using Equations 2.5 and 2.6, or the deflator for each month can be calculated by Equation 2.9 and Equation 2.10 used for the constant dollar calculation. A value for n in Equation 2.7 must first be determined for each month in either approach. Since t_0 is January 1 and the cash flows occur at the end of each month, n is simply the number of the month. The deflator for each month is then

Month	Deflator $(\bar{\$}/\$)$
January	$\dfrac{1}{[1.015]^1} = 0.985$
February	$\dfrac{1}{[1.015]^2} = 0.971$
March	$\dfrac{1}{[1.015]^3} = 0.956$
April	$\dfrac{1}{[1.015]^4} = 0.942$
May	$\dfrac{1}{[1.015]^5} = 0.928$
June	$\dfrac{1}{[1.015]^6} = 0.915$
July	$\dfrac{1}{[1.015]^7} = 0.901$
August	$\dfrac{1}{[1.015]^8} = 0.888$
September	$\dfrac{1}{[1.015]^9} = 0.875$
October	$\dfrac{1}{[1.015]^{10}} = 0.862$
November	$\dfrac{1}{[1.015]^{11}} = 0.849$
December	$\dfrac{1}{[1.015]^{12}} = 0.836$

Now the value of each cash flow can be expressed in terms of the January 1 dollar by multiplying it by its respective deflator:

Month	Dollar Cash Flow	Value Flow t_0 = January 1
January	$750,000 × 0.985 $̄/$	$̄739,000
February	775,000 × 0.971	753,000
March	795,000 × 0.956	760,000
April	800,000 × 0.942	754,000
May	800,000 × 0.928	742,000
June	800,000 × 0.915	732,000
July	800,000 × 0.901	721,000
August	810,000 × 0.888	719,000
September	810,000 × 0.875	709,000
October	800,000 × 0.862	690,000
November	780,000 × 0.849	662,000
December	760,000 × 0.836	635,000

The constant dollar, or value, cash flows are substantially different from the dollar cash flows. An analysis using the dollar cash flows could easily misrepresent the true value being exchanged. Note that the symbol $̄ is used for the cash flows when they are expressed in value. The symbol $ is used when they are expressed in dollars. Note also that the deflator has units of $̄/$.

Example 2.3

Assuming that the price indexes are a reasonable representation of the value of the dollar, determine the annual inflation rates for the years 1975–1979 using the WPI, the CPI, and the GNP-IPD.

The relationship in Equation 2.14 can be applied directly to the price indexes in Table 2.1 since they are stated on an annual basis and the Δt stated in the problem is for one year. The results of the calculations are as follows:

Time Period	Annual Inflation Rate		
	WPI	CPI	GNP-IPD
1975–1976	$\dfrac{183.0}{174.9} - 1 = 4.6\%$	$\dfrac{170.5}{161.2} - 1 = 5.8\%$	$\dfrac{169.2}{161.0} - 1 = 5.1\%$

$$1976\text{--}1977 \quad \frac{194.2}{183.0} - 1 = 6.1\% \qquad \frac{181.5}{170.5} - 1 = 6.5\% \qquad \frac{179.4}{169.2} - 1 = 6.0\%$$

$$1977\text{--}1978 \quad \frac{209.3}{194.2} - 1 = 7.8\% \qquad \frac{195.4}{181.5} - 1 = 7.7\% \qquad \frac{192.4}{179.4} - 1 = 7.2\%$$

$$1978\text{--}1979 \quad \frac{235.4}{209.3} - 1 = 12.5\% \qquad \frac{217.4}{195.4} - 1 = 11.3\% \qquad \frac{209.6}{192.4} - 1 = 8.9\%$$

It is seen that there is significant, but not great, difference between the inflation rates given by the three price indexes. Thus there is some room for disagreement as to the "true" inflation rate. These differences are not of great importance to most engineering economics problems. Engineering economics generally deals with the future where no values of the indexes yet exist anyway.

Example 2.4

A small project resulted in the dollar cash flows tabulated in the following list. Express these cash flows in value in terms of 1975 dollars, using the GNP-IPD as a measure of inflation:

Year	Revenue ($)
1970	1050
1971	1075
1972	1100
1973	1150
1974	1200
1975	1250
1976	1300

Expressing these in constant dollar terms is straightforward. The deflator for each year can be calculated from the price indexes, with 1975 as the specified reference year. The deflator is then the ratio of the 1975 GNP-IPD to the value of this index for a given year. The constant dollar, or value, cash flow is found by multiplying the dollar cash flow by the deflator:

Year	Deflator ($\bar{\$}/\$$)	Cash Flow ($)	Value Flow (in 1975 Dollars) ($)
1970	$\dfrac{161.0}{115.7} = 1.39$	1050	1460
1971	$\dfrac{161.0}{121.5} = 1.33$	1075	1420
1972	$\dfrac{161.0}{126.6} = 1.27$	1100	1400
1973	$\dfrac{161.0}{133.9} = 1.20$	1150	1380
1974	$\dfrac{161.0}{146.8} = 1.10$	1200	1320
1975	$\dfrac{161.0}{161.0} = 1.00$	1250	1250
1976	$\dfrac{161.0}{169.2} = 0.95$	1300	1240

Instead of a steadily increasing income, as the dollar cash flows indicate, the project provided a steadily declining income when expressed in terms of the actual value represented by the cash flows. The relative results would be the same regardless of the reference year chosen. Changing the reference would only mean that the "161.0" used in calculating all the deflators would be some other value. The result would be the same as multiplying all the deflators, and hence the value flows, by a constant.

The term "deflator" might not seem totally appropriate since the 1970–1974 cash flows were increased rather than decreased to express them in constant dollars. It might seem more logical to call the adjustment factor an *inflator*. However, the deflator here is defined as the conversion factor between dollars and some measure of value. No concern is given as to whether it results in an increase or decrease in numerical value.

Example 2.5

A study was made to determine the cost of modifying a plant air conditioning system in 1969. The modification consisted of changing the piping and valving to allow more control and flexibility in operating the system. The material cost estimates were as follows in 1969:

Copper tubing and fittings	$3300
Valves	850
Construction materials	1200
Miscellaneous	750
Total materials	$6100

The modification was never made but is now (1979) being reconsidered. A quick estimate of the materials cost is again needed to make a new cost estimate.

The 1969 estimate is of little direct value since prices increased substantially between 1969 and 1979. Using these cost estimates would be very misleading. A simple method to update this earlier estimate is to assume that the prices of these items have increased at a rate similar to all other items. That is, assume that the change in one of the general indexes is indicative of how these prices have changed. The WPI is probably the best index for this estimate since these items are more typical of items in the WPI than the CPI or goods and services in general as measured by the GNP-IPD. This index selection is still fairly arbitrary, however.

The general WPI changed from 106.5 in 1969 to 235.4 in 1979. This change indicates that wholesale prices increased by the factor

$$\frac{235.4}{106.5} = 2.21$$

during this time period. Therefore, a rough estimate of the change in the total cost of the material is to assume that it increased by this ratio also. The new cost estimate is then

$$\$6100 \times 2.21 = \$13,500$$

A more detailed, and hopefully more accurate, estimate can be made by updating the cost of each type of material separately. The prices of different items do not all change by the same amount. A finer breakdown of the costs and price changes should provide better insight into the present-day cost of this project.

First, each material category must be matched to a WPI category. Very detailed data are available from the source listed in Table 2.1. Appropriate matchings might be as follows:

Material Category	WPI Category
Copper tubing and fittings	Copper products
Valves	Industrial valves
Construction materials	Construction materials
Miscellaneous	Overall index

These matchings may be fairly arbitrary in some cases, but generally it is possible to find the most suitable category for a given item.

Now it is necessary to determine how much the prices have changed in each category. This information can be determined from the disaggregated price indexes available from the same source. The original cost estimate for each item is then multiplied by the 1979:1969 price ratio to arrive at a revised estimate:

Material Category	WPI Category	1969 Price Index	1979 Price Index	Price Ratio	1969 Estimate ($)	1979 Estimate ($)
Copper tubing and fittings	Copper products	105.4	188.5	1.79	3,300	5,900
Valves	Industrial valves	107.3	254.0	2.37	850	2,010
Construction materials	Construction material	106.7	230.8	2.16	1,200	2,600
Miscellaneous	Overall	106.5	235.4	2.21	750	1,660
				Totals	6,100	12,200

This estimate is somewhat lower at $12,200 than the simple estimate at $13,500. This lower estimate reflects the relatively moderate increase in copper products—a major component of cost—over this time period. It should be realized that both are rough estimates and are not substitutes for making new cost calculations. Presumably time was not available to make a new cost study in this case as a "quick" estimate was requested. The important point to be made is that cost estimates, the ultimate basis of sound engineering economic analyses, can very quickly become invalid during periods of significant inflation. Any cost estimate a few years old, or even a few months old if inflation is high, should be regarded with suspicion as to its current accuracy. The price changes are particularly a problem where more than one alternative is involved. If one alternative is judged on old cost data and another is judged on current costs, a very unfair comparison may result. It is very important to use current information whenever possible. If old data must be used, then adjustments as in this example should be made to at least recognize the fact that inflation does exist.

Example 2.6

A trucking company's study on fuel cost makes the following forecast of future diesel fuel prices:

Year	1980	1981	1982	1983	1984	1985
Diesel Price ($/gal)	1.05	1.15	1.25	1.30	1.35	1.40

The same study also forecasts a 10% inflation rate over this time period. What "real" price change in diesel fuel will occur if both of these forecasts prove to be correct?

The real price change can be found by direct application of Equation 2.16. It is more convenient to replace the price ratio with the deflator, however. If we let 1980 be the reference year, the constant dollar costs of the diesel fuel are as follows:

Year	Price ($/gal)	Deflator ($/$)	Price ($/gal)
1980	1.05	$\dfrac{1}{[1.0 + 0.1]^0} = 1.0000$	1.05
1981	1.15	$\dfrac{1}{[1.0 + 0.1]^1} = 0.9091$	1.05
1982	1.25	$\dfrac{1}{[1.0 + 0.1]^2} = 0.8264$	1.03
1983	1.30	$\dfrac{1}{[1.0 + 0.1]^3} = 0.7513$	0.98
1984	1.35	$\dfrac{1}{[1.0 + 0.1]^4} = 0.6830$	0.92
1985	1.40	$\dfrac{1}{[1.0 + 0.1]^5} = 0.6209$	0.87

When measured in constant value dollars, the price of diesel fuel decreases, assuming both forecasts to be correct. That is, the diesel fuel is expected to increase less rapidly than prices in general.

Example 2.7

A particular feed processing plant is going to need major maintenance. This maintenance will cost $250,000 to do now but could be postposed for up to two years. The cost of having this particular type of maintenance done has been increasing about 10% each year and is expected to continue to do so. The plant engineer suggests that the maintenance be done now to avoid this increased cost. The company projects the inflation rate to be about 15% for the next two years. On the basis of cost and inflation projections, how much will be saved by doing the maintenance now?

The first response by most people to this problem, and evidently the plant engineer's response, is to determine the dollar cost of doing the maintenance two years from now and to compare this cost to the cost of doing it now. If the cost projection is correct, the cost will increase 10% next year to

$$\$250,000 \times [1.0 + 0.1] = \$275,000$$

and then increase another 10% the following year to

$$\$275,000 \times [1.0 + 0.1] = \$303,000$$

It would appear that the net savings of doing the maintenance now as compared to doing it in two years would be

$$\$303,000 - \$250,000 = \$53,000$$

a very substantial apparent savings. However, the $250,000 expense of doing it now is not measured in the same value units as the $303,000 expense of doing it in two years since the value of the dollar is expected to change. Comparing these two numbers is analogous to comparing two distances, one measure in feet 12 inches long and the other measure in feet 10 inches long. A valid comparison requires that both expenses be compared in the same units of value. Both expenses should be expressed in a constant value dollar.

A constant dollar comparison requires a reference time to be selected. Any reasonable time may be used, but the present is the most convenient reference time for this problem. Using the present as a reference, the deflator for today is simply 1.0 and the deflator for two years from now, based on the

15% inflation rate, is

$$D(2 \text{ years}) = \frac{1}{[1 + 0.15]^2} = 0.76$$

The resulting costs, in "today's dollar," of doing the maintenance at the different times are as follows:

Cost today $250,000 \times 1.0$ \$/\$ = \$250,000
Cost in two years $303,000 \times 0.75$ \$/\$ = \$230,000

It actually costs less, not more, to do the maintenance in two years when the expenses are compared in consistent units.

Example 2.8*

A government agency reported that the inflation rate for a particular month was 1.2%. What annual inflation rate would be equivalent to this monthly rate?

This problem may be solved by direct application of Equation A2.1-4. The annual inflation rate is f_2 in this equation and the monthly inflation rate is f_1. The equivalent annual inflation rate is then

$$f_2 = [1 + 0.012]^{[1 \text{ yr}/1/12 \text{ yr}]} - 1 = 0.154 = 15.4\%$$

Note that the percentage inflation rate had to be expressed as a fraction in order to apply to Equation A2.1-4. Also note that care had to be used to be sure that the terms in the ratio $\Delta t_1 / \Delta t_2$ were expressed in the same units.

REFERENCES

1. *Business Conditions Digest*, U.S. Department of Commerce, Bureau of Economic Analysis, June 1969, p. 110.
2. Ibid., May 1969, p. 111.
3. Ibid., October 1969, p. 107.

APPENDIX 2.1   EQUIVALENT INFLATION RATES

Two parameters are required to completely define an inflation rate: the parameter f, which is the inflation rate, and the parameter Δt, which is the

*This example pertains to Appendix 2.1.

length of time f applies to. An inflation rate that does not specify Δt is meaningless. However, a Δt of 1 yr is usually implied when it is not specified. It is often desirable to convert an inflation rate specified for one length of time to an equivalent inflation rate for a different length of time. Equations 2.3 and 2.4 can be used to develop a relationship to determine such equivalent rates.

Let Δt_1 and Δt_2 be the time period for inflation rates f_1 and f_2, respectively. If f_1 and f_2 are equivalent, they should result in the same change in the value represented by a dollar for some given elasped time $[t - t_0]$. Then by Equation 2.7

$$n_1 = \frac{t - t_0}{\Delta t_1} \tag{A2.1-1}$$

and

$$n_2 = \frac{t - t_0}{\Delta t_2} \tag{A2.1-2}$$

Combining Equation A2.1-1 and A2.1-2 with Equation 2.6 and our statement of equivalence, it follows that

$$[1 + f_1]^{[(t - t_0)/\Delta t_1]} = [1 + f_2]^{[(t - t_0)/\Delta t_2]} \tag{A2.1-3}$$

Simplifying and reducing Equation A2.1-3 gives our final result:

$$f_2 = [1 + f_1]^{[\Delta t_2/\Delta t_1]} - 1 \tag{A2.1-4}$$

APPENDIX 2.2   COST INFLATION AND VALUE

Defining material value is a difficult and elusive goal. We are so used to defining value in terms of dollars that other measures are difficult to comprehend. The reliability of a dollar as a standard of material value is seen to be relatively poor. The working definition used in the text bases value on goods and services. The term "real price" or "real cost" is used to reflect the price of a good or service relative to the price of other goods and services. This approach is generally satisfactory for engineering economic studies; however, some subtle aspects of value and inflation are overlooked.

A somewhat different approach toward value is taken in this appendix. Consider the simple supply-demand representation of an economy shown in Figure A2.2-1. The economy produces goods and services (shown on the abscissa) using labor and other resources (shown on the ordinate). Real cost in this system is the labor and resources the people must provide to produce goods and services. Thus the ordinate is a measure of real cost. The demand curve D is a representation of the value people place on goods and services. It shows how much of their labor and resources they are willing to "pay" for a given amount of goods and services. The supply curve S is a representation of the productive capability of the economy. It shows the input of labor and resources required to produce a given amount of goods and services. The intersection (X, Y) of the two curves denotes where supply and demand are in balance. The real price of goods and services in this situation is the slope of line p.

No unit of money is needed for this supply-demand representation. Any price changes that result from shifts in D or S are real price changes in that changes in the amount of labor and resources required per unit of goods and services result. Now if we assume that price equals cost in this simple

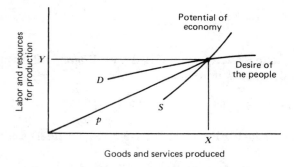

Figure A2-2.1   Simple supply-demand representation of an economy.

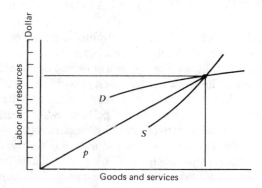

Figure A2-2.2   Introduction of money into the representation.

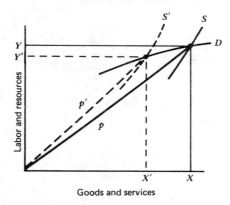

Figure A2-2.3   Effect of cost inflation on the real price of goods and services.

economy, dollar units can be introduced into the graphical representation. A dollar scale can be superimposed with the real scale on the ordinate as in Figure A2.2-2. The price of goods and services, which is the slope of p, may be evaluated either in real terms or dollar terms. The relative size of the dollar scale and real scale on the ordinate determine the real value of a dollar.

Pure price inflation with no disallocation effects will cause the dollar scale to contract while everything else in Figure A2.2-2 remains the same. Price inflation results in a decrease in the real value of the dollar but does not affect the real price of goods and services. Cost inflation results in a shift of the supply curve upward and to the left as shown in Figure A2.2-3. At the new equilibrium point (X', Y') there is a real price increase for goods and services as the slope of p' is greater than p. Even though there is no change in the real value of the dollar, there is an increase in the dollar price of goods and services with cost inflation. The value people place on goods and services may also change, resulting in a shift in the demand curve as shown in Figure A2.2-4. Again, there will be a change in the dollar price of goods and services even though the real value of the dollar does not change.

The important point of this argument is that changes in the price of goods and services does not automatically mean that the real value of the dollar has changed. Evaluating the dollar strictly in terms of goods and services may distort its true real value. Unfortunately, it is difficult to separate real price changes from dollar price changes in the real world. The difference is clear in theory but becomes blurred in reality. These problems may seem to limit the usefulness of the inflation corrections presented in the text. However, it will be seen that these inconsistencies are problematic more in theory than in practice.

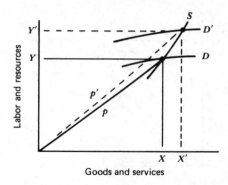

Figure A2-2.4   Effect of an increase in the value placed on goods and services.

The theory in this appendix is presented from a macroscopic point of view: the whole economy. Engineering economics deal with the point of view of an individual or a firm. The material value of a dollar to the individual is judged in terms of what it can buy and not necessarily the amount of labor and resources of other people that it represents. Furthermore, the demand curves presented here are an aggregation of the value of goods to all people. A particular individual does not necessarily have the same perception of value. Cost inflation in the economy that might shift the supply curve also might not be experienced by a particular individual. Thus a dollar value based on goods and services is probably more useful for engineering economics that a "theoretically correct" value even if one could be readily calculated.

CHAPTER THREE

The Time Value of Money

All students of engineering economics should be familiar with the concept of the time value of money. It is the central point of basic engineering economics theory. It is from the concept of time value that the mechanics of engineering economic analysis are derived. Consequently, the effect of inflation on time value is the starting point for determining how inflation affects engineering economic analysis. Before studying inflation further, it is necessary to review the meaning of the time value of money and why it exists.

3.1 THE VALUE OF WEALTH

As was stated earlier, money (the dollar) has no intrinsic value. Its value is due to its ability to buy valuable goods and services. Money also has no time value unless the goods and services it buys have a time value. Thus the term "time value of money" is rather misleading. It might be more correct to refer to the "time value of wealth," meaning the time value of goods and services. Money has no time value unless the "wealth" it can be exchanged for has a time value.

The time value of wealth must be examined from two points of view. The first view deals with the material aspects of time value. It deals with the ability of goods and services to earn a return. The second view deals with the psychology of time value. It deals with the relative value people place on wealth as a function of time.

To review the fundamental meaning of time value, it helps to use the simple macroscopic overview of an economy shown in Figure 3.1. Labor and resources are combined through the capital stock of the economy to produce goods and services. Capital stock includes factories, power plants, machinery, farms, offices, and so on. It is the capital stock of an economy that provides the means for labor and natural resources to be converted into

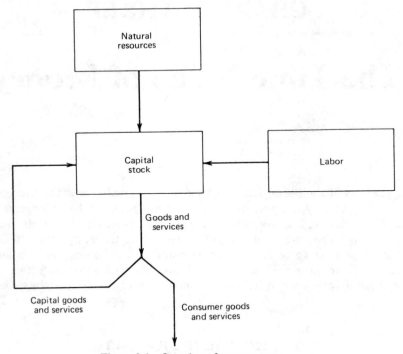

Figure 3.1   Overview of an economy.

goods and services. In a general sense, capital stock includes less tangible products like education, scientific knowledge, and social systems. Goods and services produced in the economy can be divided into two different, but not totally distinct, groups as shown in Figure 3.1. Consumer goods and services are used by the people. They are the items produced for the people to "consume." Consumer goods and services include clothes, food, health services, housing, recreation facilities, and so on. Capital goods and services are produced to add to the capital stock. They will be used in producing goods and services in the future.

Presumably, the purpose of an economy is to produce goods and services for the people. Diverting the output of the economy from consumer items to capital items would not be logical unless the capital items increased the ultimate output of consumer goods and services. Very little in the way of goods and services can be produced directly from natural resources and labor. Only very primitive economies function without a significant capital stock. It is the ability of capital stock to increase the production of goods and services that gives rise to a time value of wealth. Suppose that some quantity of wealth X existing as capital stock increases the output of the economy by amount ΔX per unit of time Δt. Then this wealth may be said to earn a return R as follows*:

$$R = \frac{\frac{\Delta X}{\Delta t}}{X} \tag{3.1}$$

The output of the economy diverted to capital items increases the rate at which the economy produces goods and services as a function of this return:

$$\Delta O = \frac{\Delta X}{\Delta t} = R \cdot X \tag{3.2}$$

where ΔO is the increase in the rate at which goods and services are produced in the economy.

Not all capital stock earns the same return. At any time there are limited numbers of opportunities to use capital stock. Some will yield a high return, others a lower return. If all the opportunities to use capital stock are arranged in order according to the return they yield, a curve as in Figure 3.2 can be developed. Increased production of capital stock results in relatively less increase in the output of the economy.

What, then, determines how much production will be diverted to capital stock? How much return is required to make this diversion worthwhile? The answers to these questions are determined by the relative value placed on consumption now as compared to consumption in the future. All else being equal, most people would rather have consumer items now than the promise of the same items in the future. The relative value now of goods and services now is greater than the value now of the same goods and services at some time in the future. These relative values are based on human perception and cannot necessarily be precisely defined. However, it may be postulated that increasing return on output diverted to capital stock is required to increase the amount diverted. The result is a curve as shown in Figure 3.3. As more and more consumption is foregone to produce capital items, increased production is required to make this diversion worthwhile.

Some process is required to combine the opportunities for capital investment shown in Figure 3.2 with the relative time value placed on consumption represented in Figure 3.3. The required decisions may be made individually, collectively, or by other means. However they are made, the fact remains

*The question of the productive life of the capital stock is ignored for the present discussion.

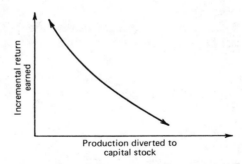

Figure 3.2   Return earned on capital stock.

that whatever output is diverted to capital stock represents foregone consumption. It is tempting to combine the curves in Figures 3.2 and 3.3 as representations of investment supply and demand in an idealized free economy as shown in Figure 3.4. The willingness to forego present consumption so that output may be diverted to capital stock represents supply of investment. The opportunities to use capital stock to increase output represent demand for investment. Such an approach is certainly an oversimplification of the real world. Government policies, improper wealth distribution, ignorance of investment opportunities, and so on may interfere with this simple matching of supply and demand.

If it is assumed that an ideal capital investment market as in Figure 3.4 exists, then the equilibrium point (R', X') determines the amount of output diverted to capital stock and the return that can be earned on additional investments. The capital investment market appears as an infinite set of investment possibilities that will earn a return R' to the small investor. Any amount of wealth now may be invested through the capital market in

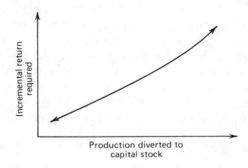

Figure 3.3   Return required to induce production of capital items.

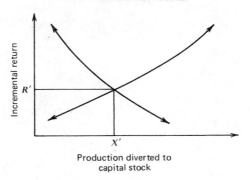

Figure 3.4 Capital investment supply and demand.

exchange for increased wealth in the future according to the market equilibrium return:

$$\frac{\Delta X}{\Delta t} = X \cdot R' \tag{3.3}$$

where $(\Delta X/\Delta t)$ is now the increased rate at which the investor receives wealth and X is the amount of wealth the investor places into the capital market now. Similarly, the capital investment market may be viewed as a source of wealth. An individual may receive wealth of amount X now but must return wealth according to Equation 3.3.

The capital market provides a mechanism for transferring wealth from one time to another time. Assuming that wealth may be freely put into or taken out of the market, an amount of wealth today X may be exchanged for an amount of wealth $X + \Delta X$ after some time Δt:

$$X + \Delta X = X + X \cdot R' \cdot \Delta t \tag{3.4}$$

The value of wealth may then be said to be a function of time according to Equation: 3.4:

$$V(X, \Delta t) = V(X, 0) \cdot [1 + R'] \tag{3.5}$$

where $V(X,0)$ is the value placed now on wealth of quantity X and $V(X,\Delta t)$ is the value of wealth it will provide after some time Δt. The value of wealth $V(X, \Delta t)$ may be withdrawn and then reinvested in the capital market at time Δt. It will further increase in value during this second time period:

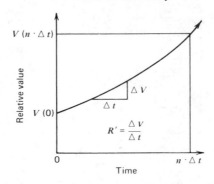

Figure 3.5 Change in the value of wealth with time.

$$V(X, 2 \cdot \Delta t) = V(X, \Delta t) \cdot [1 + R'] \qquad (3.6)$$

These processes may be extended to any number of time periods. The result is a function as shown in Figure 3.5 that describes the relative value of wealth as a function of time. If it is assumed that R' stays constant, the function may be expressed in terms of simple mathematics:

$$V(X, n \cdot \Delta t) = V(X, 0) \cdot [1 + R']^n \qquad (3.7)$$

The curve shown in Figure 3.5, defined by the capital market, expresses the time value of wealth. It shows how much wealth at one point in time may (or must) be exchanged for an amount of wealth at another point in time. Any two points on the curve should represent wealth of equal value to the individual. When evaluating economic opportunities, the individual would be indifferent to an income of $V(X, n_1 \cdot \Delta t)$ at time $n_1 \cdot \Delta t$ or an income of $V(X, n_2 \cdot \Delta t)$ at time $n_2 \cdot \Delta t$. Even if for some reason the individual preferred one over the other, they could be exchanged for each other through the capital market. This concept of indifference is another way of looking at the time value of wealth R' and is commonly referred to as equivalence. The time value of wealth results in different quantities of wealth at different times being equivalent as described by Equation 3.7 and as represented by Figure 3.5.

The real capital market is not as ideal as described here. Different people and firms will have access to different costs and different opportunities. They will experience different risks. One often cannot obtain wealth (borrow) at the same return that it may be invested in the market. The future may not be sufficiently certain to know what wealth will be provided by additional capital stock. Nevertheless, the marketplace does establish a means whereby current wealth may be exchanged for wealth at a future time. The exact time value of wealth that a particular individual experiences in the market may be difficult to determine, but its existence must not be overlooked, and an economic analysis should reflect the time value of wealth.

3.2 REAL VERSUS APPARENT TIME VALUE

The existence of a time value of wealth gives rise to a time value of money since money may be exchanged for wealth. Some quantity of money may be invested in the capital market now in exchange for a larger amount at some time in the future. Or some amount of money may be borrowed now in exchange for a larger amount at some time in the future. The return on money in the market is commonly referred to as interest and is denoted by the symbol i. Since the time value of money is based on the wealth that money can buy, i has the same meaning as R' although they may not be the same in magnitude. It should be realized, however, that money must ultimately be exchanged for capital goods, either directly or through the capital market, before it has a time value. Money that is simply held has no time value.

The expression of the time value of wealth in Equation 3.7 may be converted to dollar terms by Equation 2.3:

$$Y^d(n \cdot \Delta t) = V(X, n \cdot \Delta t) \cdot \bar{p}(n \cdot \Delta t) \qquad (3.8)$$

where Y^d is the dollar cash flow that must be exchanged for wealth of value V. Equivalent dollar cash flows are then given by

$$\frac{Y^d(n \cdot \Delta t)}{\bar{p}(n \cdot \Delta t)} = \frac{Y^d(0)}{\bar{p}(0)} [1 + R']^n \qquad (3.9)$$

where $Y^d(n \cdot \Delta t)$ and $Y^d(0)$ are different cash flows that are economically equivalent. The interest rate is defined as

$$Y^d(\Delta t) = Y^d(0) \cdot [1 + i] \qquad (3.10)$$

where i is the return earned per dollar in time period Δt. It is analogous to R'. Assuming a constant interest rate, Equation 3.10 may be expanded to any number of time periods:

$$Y^d(n \cdot \Delta t) = Y^d(0) \cdot [1 + i]^n \qquad (3.11)$$

The relationship between i and R' may be found by combining Equations 3.11 and 3.9:

$$\frac{Y^d(0) \cdot [1 + i]^n}{\bar{p}(n \cdot \Delta t)} = \frac{Y^d(0)}{\bar{p}(0)} \cdot [1 + R']^n$$

$$\frac{[1 + i]^n}{\bar{p}(n \cdot \Delta t)} = \frac{[1 + R']^n}{\bar{p}(0)}$$

$$[1 + i] = [1 + R'] \cdot \left[\frac{\bar{p}(n \cdot \Delta t)}{\bar{p}(0)}\right]^{1/n} \tag{3.12}$$

Equation 3.12 shows the fundamental relationship between the time value of money and the time value of wealth. If prices do not change, the two forms of time value are the same. Inflation results in changing prices and, consequently, differences in the two forms of time value. If the inflation rate is assumed to stay constant, the price term in Equation 3.12 may be replaced by the expression for inflation given in Equation 2.6:

$$\left[\frac{\bar{p}(n \cdot \Delta t)}{\bar{p}(0)}\right]^{1/n} = \{[1 + f]^n\}^{1/n} = 1 + f \tag{3.13}$$

Equation 3.13 then becomes

$$[1 + i] = [1 + R'] \cdot [1 + f] \tag{3.14}$$

The time value of money in dollar terms may be misleading in that it does not show the true gain in wealth. The term "apparent time value," denoted by the symbol i_a, is used hereafter when reference is made to the dollar time value of money. The time value of wealth is referred to as the "real time value" and is denoted by the symbol i_r. If these definitions are used, Equation 3.14 becomes

$$[1 + i_a] = [1 + i_r] \cdot [1 + f] \tag{3.15}$$

Dollar cash flows were converted to value representation through the constant dollar concept presented in Chapter 2. If the constant dollar representation is a valid expression of value, the time value of money expressed in constant dollar terms should be equal to the real time value in Equation 3.15. This relation can be seen by substituting constant dollars according to Equation 2.5 for the dollars in Equation 3.11:

$$Y^c(n \cdot \Delta t) \cdot \frac{\bar{p}(n \cdot \Delta t)}{\bar{p}(t_0)} = Y^c(0) \cdot \frac{\bar{p}(0)}{\bar{p}(t_0)} \cdot [1 + i_a]^n$$

$$Y^c(n \cdot \Delta t) = Y^c(0) \cdot \frac{\bar{p}(0)}{\bar{p}(n \cdot \Delta t)} \cdot [1 + i_a]^n \qquad (3.16)$$

where Y^c is the constant dollar representation of the dollar cash flows Y^d. The price term may be eliminated by again assuming a constant inflation rate and using Equation 2.6:

$$Y^c(n \cdot \Delta t) = Y^c(0) \cdot \frac{[1 + i_a]^n}{[1 + f]^n} \qquad (3.17)$$

By Equation 3.15, this expression may be reduced to

$$Y^c(n \cdot \Delta t) = Y^c(0) \cdot [1 + i_r]^n \qquad (3.18)$$

It is seen then that expressing equivalence in constant dollars and real time value is the same as expressing equivalence in dollars and the apparent time value:

$$Y^d(n \cdot \Delta t) = Y^d(0) \cdot [1 + i_a]^n \qquad (3.19)$$

The relationship between the real and apparent time value is due to inflation as shown in Equation 3.15. It should be remembered, however, that these expressions were derived with assumptions of constant time values and constant inflation. The science of engineering economics is sufficiently imprecise that these assumptions are usually justified. The more difficult situation of varying inflation and time value is discussed in Chapter 11.

3.3   THE INFLATION COMPONENT OF TIME VALUE

The money market operates in terms of dollars. Loans, interest payments, contracts, dividends, and so on are specified in terms of dollars without regard to the value of those dollars. Traditional interest rates are thus apparent interest rates and not real interest rates. Since people are so used to working in dollar terms and the money market works in dollar terms, it is traditional to express the time value of money in dollar terms also. This tradition does not mean that investors are ignorant of the difference between real and apparent returns on investments, nor does it mean that the money market does not respond to changes in the value of the dollar. It simply means that these considerations are reflected in the dollar time value assigned to money.

The time value of money, as it is reflected in the capital market, is often referred to as the *cost of money*. The cost of money in its traditional form can actually be separated into three components:

1. A cost reflecting the real time value of wealth.
2. A cost reflecting the decreasing value of the dollar.
3. A cost reflecting the risk in an investment.

It is to our advantage to view each of these elements separately.

Although cost item 3 is usually included in the time value of money, it can be considered separately. The general theories for evaluating risks involve complicated statistical analysis. However, a simple example can demonstrate the basic principle.

Consider a situation where the time value of money, exclusive of risks, is reflected in a 5% annual interest rate. An investment with a one year life would have to return at least 5% more than the original amount expended to be worthwhile. Now consider investments with a life of one year that fail about 20% of the time. That is, 20% of these investments would return nothing. If a large number of these investments were made, the return after one year would be

$$Y_r = Y_i \cdot [1 + r] \cdot 0.80$$

where Y_r is the money returned, Y_i is the original amount of money invested, r is the typical return for an investment of this type that does not fail, and the 0.80 is due to the fact that 20% of these investments return nothing. For the net return on these investments to be 5%, Y_r must be 5% larger than Y_i:

$$Y_r = Y_i \cdot [1.05]$$

These two equations can be solved for r:

$$Y_i \cdot [1.05] = Y_i \cdot [1 + r] \cdot 0.80$$

$$[1.0 + r] = \frac{1.05}{0.80}$$

$$r = 0.3125$$

$$r \approx 31\%$$

These investments would then need to earn a 31% return to be worthwhile. Investors might say that money had a time value, or cost, of 31% for these particular investment even though it only has a 5% cost for risk free investments.

A theoretically more sound approach would be to leave the time value of money at the risk free 5% but to evaluate the investments on their probable outcome rather than their outcome when they succeed. Suppose that one of these investments is expected to return $130 if it succeeds. This investment has only at 80% probability of success; so rather than saying that its return will be $130, we could say that the probable return will be $130 $\times$ 0.80 = $104. These investments will return $104 on the average. This return can now be evaluated using the 5% time value of money rather than the "artificial" 31% time value of money. Thus if risks are accounted for in the returns from an investment, it is not necessary to include a risk cost in the cost of money. This later approach is gaining acceptance for analysis within a company. However, the money market generally reflects the former approach. Table 3.1 shows the yield of corporate bonds as a function of their risk category. It is seen that the greater the risk, the greater the yield. This greater return is required to offset the probability that the bonds may not be repaid.

TABLE 3.1   Average Coporate Bond Yield (July 1980)[1]

	Risk Category			
	Aaa	Aa	A	Baa
Yield (%)	11.07	11.43	11.95	12.65

There are some transactions in the money market that have little risk associated with them.* Short-term U.S. Government securities or Aaa corporate bonds are such items. There is almost no chance that these "loans" will not be repaid. The market interest rate on these items should then reflect primarily the time value of wealth and the decreasing value of the dollar.

Figure 3.6 shows the yield of three month treasury bills for the last 25 years. Inflation as measured by the GNP-IPD is shown on the same graph. Also, the real return as calculated by Equation 3.14 is shown. It is seen that the yield of the treasury bills is nearly coincident with inflation. The cost of money reflected by these bills is almost entirely due to the decreasing value of the dollar rather than the time value of wealth. The real yield is relatively modest and sometimes negative.

*The term "no risk" here means that any dollar amount specified will be paid. There still may be a substantial real risk associated with such investments, as is explained in Chapter 9.

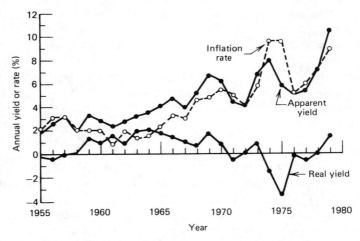

Figure 3.6 Yield of three month government securities.[2]

These 90 day treasury bills may not be the best measure of the cost of money, but they do show the strong tie between inflation and interest costs. A good deal of corporate financing is done through bond issues. Thus the yield on corporate bonds is an important measure of the cost of money for corporate investments. Class Aaa bonds have relatively little risk associated with them so their yield should not reflect any risk component of the cost of money.

Figure 3.7 presents the average yields of these bonds for the past 25 years and inflation as measured by the GNP-IPD. Again, a strong tie between the cost of money and inflation is seen. The real yield of these bonds is relatively

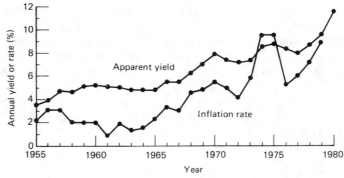

Figure 3.7 Apparent yield of Aaa corporate bonds.[3]

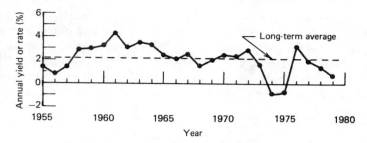

Figure 3.8   Real yield of Aaa corporate bonds.

modest. The high inflation rates of recent years have resulted in money costs that are high. However, these high interest rates are misleading in that they do not reflect increases in the real time value of wealth.

Estimating the real time value of wealth is difficult. The data in Figure 3.7 are the result of an open maket that is not necessarily in balance at any give time. Over the long term, however, it would be expected that the market would reflect the real time value of wealth. The real yields of the Aaa bonds are shown in Figure 3.8. The average yield over the 25 year period is about 2%. If the yearly data are smoothed to eliminate the short-term market fluctuations, it appears that there have been no great variations from this long-term average.

Corporate bonds are only one source for financing capital stock. Other sources may indicate a somewhat higher, or lower, real time value of wealth. However, bonds are a significant portion of the total capital financing in the United States and should at least be indicative of the real time value that capital investment earns. The real return of a few percent compared to the apparent return of 10%, 15%, and more for many bonds shows that the traditional time value of money has been more an accounting of inflation and risk rather than a reflection of the true earning power of capital stock in recent years. This situation does not necessarily lead to incorrect results in engineering economic analysis if the apparent time value is correctly used. The results of an analysis may be misleading if not carefully interpreted, and special care must be taken to be sure that analysis is set up correctly. These problems are examined in Chapter 4.

3.4   DETERMINING THE TIME VALUE OF MONEY

The preceding discussion has dealt with the time value of money as it is reflected in the marketplace. If the market were perfect and ideal, it would properly represent the time value to use in an economic analysis. The market is imperfect and is not ideal. There is no single time value of money that can be applied to all problems. Determining the correct time value to use for a particular analysis is difficult and may require some subjective judgment. The details of this process are not discussed here. However, it is important to realize that the way in which the time value is determined will control how it must be used.

Usually, the time value of money is regarded as the cost of money. A corporation may determine its cost of money by weighting the interest cost associated with its various sources, such as bond issues, retained earnings, and stock. If the cost assigned to each of these sources is in dollar terms, as is usual, the resulting time value of money is an apparent time value including the cost of inflation.

Sometimes the time value of money for a corporation or individual is based on an opportunity cost. The cost of money is said to be determined by the return from available but unused investments. That is, the time value is based on the available opportunities to earn a return. The form of the time value here depends on how the investment opportunities are viewed. If they are expressed in real, or constant, dollar terms, the resulting time value will be a real time value; if expressed in dollars, the resulting time value will be an apparent time value.

It is not always clear how a time value is expressed, so care must be exercised when using it in an analysis as the two forms of time value mean two completely different things. One is a measure of the real return from capital stock; one is primarily a correction for inflation. The apparent form is much more common and leads to correct results in most cases if used correctly. However, it tends to obscure the economic evaluation with its inflation related component. It is desirable in many problems to separate the real time value from the inflation and consider each explicitly.

3.5   EXAMPLES

Example 3.1

A small country has a simple, self-sufficient agricultural economy. The main product is wheat, which is also the dietary staple. The country's primary resource is its land, which is worked by hand. Many years of poor agricultural practices have resulted in a depleted soil that is not very productive. Annual wheat production is typically 1500 lb/acre. The farmers have found that the best way to increase production is to sow some special cover crops rather than wheat that are then tilled into the soil. This revitaliztion of the soil, along with better farming practices, is found to increase annual production by about 120 lb/acre if done for one growing season (1 yr). The productivity is increased more, but proportionately less, if the revitalization is continued for several years. The labor required to grow and till the cover crop is comparable to the requirements for raising wheat. Very little of this revitalization has been done yet. Estimate the time value of wealth in this economy.

Since wheat is the main product of the economy, it should be a good measure of wealth. The time value of wealth in the economy depends on the return possible from capital stock and the willingness of the people to forego present consumption, as was shown in Figure 3.4. The best return from capital stock is said to be from the soil revitalization. One year's production of wheat (1500 lb/acre) must be foregone to revitalize the soil. This investment returns an annual increase in production of 120 lb/acre. The return for this investment in capital stock is then

$$R = \frac{120 \text{ lb/acre}}{1500 \text{ lb/acre}} = 0.08$$

Other ways of improving yield undoubtedly exist, but it was stated that this method was the best. It would be expected that these other forms of investment would not be made as long as opportunity for the revitalization existed. The demand for investment in the economy could be represented by a straight line as in Figure 3.9. Since very little revitalization has been done,

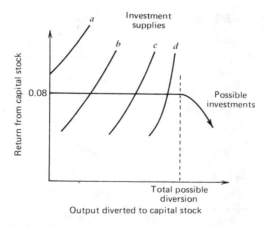

Figure 3.9   Possible investment supply-demand balances.

essentially all the productive capability of the economy could be diverted to this form of investment. Greater amounts of investment are not possible.

The supply of investment depends on how much consumption the people are willing to forego. Several possibilities are shown in Figure 3.9. The equilibrium point gives a time value of wealth of 0.08 unless curve a describes the supply of investment. In that case, there is no investment or time value. It is probably safe to say that the time value of wealth in this economy at the present time is 0.08.

Example 3.2

A particular company uses the general price index forecasts tabulated in the following list. If the company has a real time value of wealth of 4%, what dollar cash flow in 1990 would be economically equivalent to a $10,000 cash flow in 1985?

Year	1980	1981	1982	1983	1984	1985	1986	1987	1988	1989	1990
Price index	100	111	122	131	141	151	162	173	185	198	212

The principle of equivalence states that the ratio between constant dollar cash flows of equivalent economic value in 1985 and 1990 would be

$$Y^c(1990) = Y^c(1985) \cdot [1 + i_r]^5$$

or

$$Y^c(1990) = Y^c(1985) \cdot [1.04]^5$$

The dollar cash flows can be related to the constant dollar cash flows with the price index:

$$Y^c(1990) = Y^d(1990)\frac{I(t_0)}{I(1990)}$$

$$Y^c(1985) = Y^d(1985)\frac{I(t_0)}{I(1985)}$$

The equivalence statement can now be expressed in dollar cash flows rather than constant dollar cash flows:

$$Y^d(1990)\frac{I(t_0)}{I(1990)} = Y^d(1985)\frac{I(t_0)}{I(1985)} \cdot [1.04]^5$$

$$Y^d(1990) = Y^d(1985)\frac{I(1990)}{I(1985)} \cdot [1.04]^5$$

Now the price index values and the $10,000 cash flow can be substituted into the relationship to determine the equivalent dollar cash flow in 1990:

$$Y^d(1990) = \$10,000\,\frac{212}{151} \cdot [1.04]^5 = \$17,080$$

A dollar cash flow of $17,080 in 1990 is economically equivalent to $10,000 in 1985.

Example 3.3

A company with a real time value of wealth of 3.5% has made a contract that requires a cash flow of $23,500 in 1987. They expect inflation to be about 9% between now and then. What cash flow now (1980) is equivalent to the contracted cash flow?

Since the $23,500 is specified in a contract, we can be reasonably sure that it is a dollar amount. The inflation rate is given only as a constant value, so it is possible to convert the real time value to an apparent time value that is more suitable for dealing with dollar flows:

$$[1 + i_a] = [1 + i_r] \cdot [1 + f]$$
$$1 + i_a = 1.035 \cdot 1.09 = 1.1282$$
$$i_a = 12.82\%$$

The equivalence statement in dollar terms is

$$Y^d(1987) = Y^d(1980) \cdot [1 + i_a]^7$$
$$\$23,500 = Y^d(1980) \cdot [1.1282]^7$$
$$Y^d(1980) = \$10,100$$

The $23,500 cash flow that will occur in 1987 is economically equivalent to a $10,100 cash flow today.

Example 3.4

An engineer is assigned an economic analysis task. In completing the task, the engineer determines that money is available in the market to fund the project being analyzed for about 14% interest. Inflation at that time is about 9%. What real and apparent time values of money would be appropriate for the analysis?

The cost of money in the market is an appropriate time value of money. The interest cost would normally be stated in dollars, so the apparent time value should be 14%. The real time value can be evaluated by Equation 3.15:

$$1 + i_r = \frac{[1 + 0.14]}{[1 + 0.09]} = 1.046$$

$$i_r = 4.6\%$$

The real time value would be about 4.6%.

Note that the relationships between i_r and i_a are based on frational forms of interest and inflation and not percentage forms.

Example 3.5

An engineer works for a small firm that seldom borrows money in the capital market. The company uses funds from its own revenues for new investments. The company is considering some investments that pertain to a new product and wonder what time value of money should be used for related economic analyses. In reviewing the present cost of production equipment, the cost of production and the price the company receives for existing products, it is seen that the company earns at least 10% on investments with present products. Inflation is about 9% here. What real and apparent time value should be used?

The company evidently does not participate directly in the capital funds market, so the time value cannot be the cost in this market. However, the company may be depicted in much the same way as the overall economy. They have certain capital investments available and funds that they can invest or use for consumption expenditures. The supply-demand curves for investment within the company are shown in Figure 3.10. It is not known what the exact values of the curves are. However, it is seen that investments are made that earn 10% return and higher, so the intersection between investment supply and demand must be at about 10% (assuming that the company has been optimally expending its capital funds). If the new product

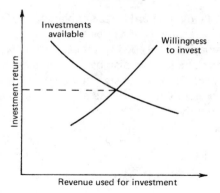

Figure 3.10 Current investment supply-demand balance within company.

investments are relatively small compared to the existing investments being considered, they should not shift the equilibrium point too much, and investments funds should have a time value of 10%. Another view of this situation is to say that there are investments available that earn a return of 10%, so investment funds can earn this return and thus have a cost of 10% when considered for other investments. Determining time value in this manner is referred to as using the "opportunity cost."

The company has a time value of 10%, but it is not clear whether this is an apparent or real time value. The 10% return was estimated using today's prices for both investment costs and net revenues. The return is then expressed in constant dollars with the present as the reference year. The 10% time value is then the real time value. The apparent time value can be found by Equation 3.15:

$$1 + i_a = [1 + 0.10] \cdot [1 + 0.09] = 1.199$$
$$i_a = 19.9\%$$

Example 3.6*

Quite frequently, people will hoard supplies when they expect war or other social disturbances to occur. This behavior is often criticized as having no productive value. The argument is that, at best, only the same amount of goods is available at a later date and hence more goods cannot result. Is there a rational explanation for such behavior?

*This example pertains to Appendix 3.1.

One does not need to look too hard for the explanation. If people anticipate that the goods will not be available at a later date and they are fairly essential items, no economic analysis is needed to explain why people hoard goods. However, a rational explanation can also be made in less extreme situations. The disturbances are likely to result in significant cost inflation, at least for a time. The true value of goods and services are likely to increase. Hoarding the goods may provide only

$$i_r = 0$$

But since

$$i_t = [1 + i_r] \cdot \frac{p(\Delta t)}{p(0)} - 1.0$$

the true rate of return will be positive if the true value of goods increases, which will result when

$$\frac{p(\Delta t)}{p(0)} > 1$$

Thus hoarding of goods in this situation is a viable form of "capital investment" in that it provides goods in the future that are of more value than the consumption foregone.

REFERENCES

1. *Moody's Bond Record*, Vol. 47, p. 2, August 1980.
2. *Ibid.*, p. 188.
3. *Ibid.*, p. 2, 182; p. 155, January 1976; p. 99, January 1973.

APPENDIX 3.1   COST INFLATION AND TIME VALUE

It was shown in Appendix 2.2 that defining inflation in terms of the price of goods and services could lead to theoretical errors due to changes in value placed on goods and services. This error is most likely to occur when there is significant cost inflation. Similar errors may occur in the time value of wealth when it is measured in terms of goods and services.

Consider a measure of time value stated in terms of the true costs of production, labor and resources. This time value is referred to as the "true" time value and denoted by the symbol i_t. The true time value describes return in labor and resources in the same way that the real time value describes return in goods and services:

$$Q(\Delta t) = Q(0) \cdot [1 + i_t] \qquad (A3.1\text{-}1)$$

where $Q(0)$ is the true value of the goods and services diverted to capital stock at time zero and $Q(\Delta t)$ is the true value of the goods and services the investment yields. The relationship between goods and services and value as measured by labor and resources was described in Appendix 2.2 and is depicted in Figure A3.1-1. The true value of goods and services in an economy is equal to the slope p of the line drawn to the intersection of the curve S describing the productive capability of the economy and the curve D describing the willingness to commit labor and resources to production. The value, or cost, of goods and services is then describe by

$$Q = p \cdot X \qquad (A3.1\text{-}2)$$

where X is some quantity of goods and services. Equation A3.1-1 can now be rewritten in terms of goods and services:

$$p(\Delta t) \cdot X(\Delta t) = p(0) \cdot X(0) \cdot [1 + i_t]$$

$$X(\Delta t) = X(0) \cdot \frac{p(0)}{p(\Delta t)} \cdot [1 + i_t] \qquad (A3.1\text{-}3)$$

The ratio

$$\frac{X(\Delta t)}{X(0)}$$

is the same as the ratio

$$\frac{Y^c(\Delta t)}{Y^c(0)}$$

since both are measures of goods and services. By Equations 3.18 and A3.1-3, the relationship between the "true" and "real" time values is

$$[1 + i_t] = \frac{p(\Delta t)}{p(0)} \cdot [1 + i_r] \qquad (A3.1\text{-}4)$$

It also follows that

$$[1 + i_t] = \frac{p(\Delta t)}{p(0)} \cdot \frac{\bar{p}(0)}{\bar{p}(\Delta t)} \cdot [1 + i_a] \qquad (A3.1\text{-}5)$$

The ratio

$$\frac{p(\Delta t)}{p(0)}$$

describes the change in the cost of goods and services in terms of the labor and resources required for their production. The ratio

$$\frac{\bar{p}(0)}{\bar{p}(\Delta t)}$$

describes the change in cost of goods and services in dollars. The relationship between the true and real time values is independent of any price inflation. The dollar scale on the vertical axis in Figure A3.1-1 can change without affecting the cost of goods and services p in terms of labor and resources. However, cost inflation will affect p as shown in Figure A3.1-1.

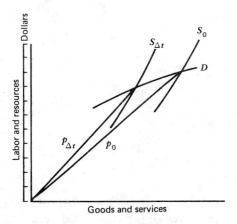

Figure A3-1.1   Changing value of goods and services due to cost inflation.

There is an important consequence of the relationship between the true and real time values. At times, negative real time values are encountered. Such a situation seems irrational. No one would want to invest today to get less goods and services back at some time in the future. However, it is quite possible that the true return is positive even though the real return is negative. Less goods and services may be returned, but they are valued more highly than the original investment. Such a situation is entirely possible during periods of extreme cost inflation.

If there is no price inflation, the relationship between the dollar scale and the labor and resources scale stays the same. Any inflation that occurs under these circumstances may be said to be pure cost inflation. The following relationship would then hold true:

$$\frac{p(\Delta t)}{p(0)} = \frac{\bar{p}(\Delta t)}{\bar{p}(0)}$$

and the apparent time value would be equal to the true time value. If there is pure price inflation, the value of p does not change, and

$$\frac{p(\Delta t)}{p(0)} = 1$$

Pure cost inflation results in the apparent time value being a measure of true time value. Pure price inflation results in the real time value being a measure of the true time value.

It could be argued that real time value in terms of goods and services may not always be the proper measure to use in economic analysis. However, these arguments pertain to a macroscopic view of an economy. Engineering economics must deal with microscopic considerations. The fortunes of an individual do not necessarily follow those of the overall economy. The cost of goods and services in labor and resources may change differently for each individual or firm. Just because there is cost inflation in the economy does not mean that a particular individual will value goods and services more highly in the future. Therefore, it is argued that the real time value is still the proper measure of investment return. It states results in terms that can be readily understood. Individuals can then decide how their particular values may change.

This conclusion does not mean that the conceptual difference between real and true time values is not important. The difference should be kept in mind when dealing with an inflationary environment. Negative real return may sometimes be rational. A negative real time value may also imply that funds are available in the capital market that require repayment with funds less costly in terms of goods and services. Such arrangements appear very attractive but offer potential pitfalls. The repayment may represent less goods and services, but these goods and services may represent more value than the original payment received.

CHAPTER FOUR

Including Inflation in Economic Analyses

4.1 DOLLAR AND VALUE TERMS

The transactions involved in the economics of investments can be described in (1) dollar terms (i.e., the actual dollar cash flows that are expected to take place) and (2) value terms (i.e., constant dollar equivalents of the actual dollar cash flows). The first, and generally the most difficult, step in an engineering economic analysis is to estimate the cash flows. We usually know the materials, labor, and other requirements and the products that may be produced, but no economic analysis can be made until items are stated in economic terms, either dollars or constant dollars. To do so, the prices of the various items must be known or estimated.

Price forecasting traditionally has been a secondary part of engineering economics. This role is understandable since the purpose of engineering economics is to aid in making wise engineering decisions. Price fluctuations create speculative problems that are not the engineer's task to solve. As much as possible, engineering economics should not get the engineer involved with such considerations but should leave the engineer to solve engineering problems. Nevertheless, engineering economics must reflect major shifts in the prices of items involved in an economic analysis. Inflation results in such shifts.

Price changes for a particular item may be divided into two parts:

$$\frac{p_j^d(t)}{p_j^d(t_0)} = \frac{\bar{p}(t)}{\bar{p}(t_0)} \cdot p_j^r \left(\frac{t}{t_0} \right) \tag{4.1}$$

The term

$$\frac{p_j^d(t)}{p_j^d(t_0)}$$

is the relative change in the dollar price of item j from time t_0 to time t. The term

$$\frac{\bar{p}(t)}{\bar{p}(t_0)}$$

is the price change of all items during the same time period and may be regarded as the inflation component of the price change. The term $p_j^r(t/t_0)$ is the relative shift in the price of item j compared to all items. It is sometimes referred to as the *real price change*. It also gives the price change in constant dollar terms:

$$\frac{p_j^c(t)}{p_j^c(t_0)} = p_j^r\left(\frac{t}{t_0}\right) \tag{4.2}$$

where p_j^c is a price stated in constant dollars.

It was seen in Chapter 3 that time value and the principle of equivalence can be stated either in dollar terms or in constant dollar value terms. If it is desired to work with real time value, all cash flows and prices should be stated in constant dollar terms. If it is desired to work with apparent time value, all cash flows and prices should be stated in dollar terms. The form of the time value determines how cash flows must be represented. It is possible to alternate between real and apparent time value with Equation 3.15 if future inflation is assumed to be constant. Such an assumption may sometimes be difficult to justify but is probably no worse than assuming that the apparent time value is constant, as is generally done in economic analysis. Using the constant inflation assumption, Equation 4.1 may be rewritten as

$$\frac{p_j^d(t)}{p_j^d(t_0)} = [1 + f]^n \cdot p_j^r\left(\frac{t}{t_0}\right) \tag{4.3}$$

where

$$n = \frac{t - t_0}{\Delta t}$$

and Δt is the time period to which f applies, normally one year.

Predicting dollar prices then requires predicting both the inflation rate and the real price change. Such an approach may seem to add complexity to the problem but actually makes the task less difficult. During periods of significant inflation, the real price changes are normally much less than changes due to inflation. It is essential that all price forecasts for a particular analysis be based on the same inflation rate. Independent price forecasts for different items may reflect different assumptions about future inflation rates and grossly overstate future price differences. Expressing price forecasts in the form of Equation 4.3 ensures that the forecasts are consistent since they are all based on the same inflation rate. Any real price shifts must be justified with marketplace or other considerations. Although it is dangerous to generalize, it is probably safe to say that for engineering economics purposes relative prices should be near unity unless there are some convincing reasons why major changes will be forthcoming in real costs of production (depleted sources of materials, new government regulations, price fixing by cartels, new technological developments, etc.)

Dollar prices must not only be consistent with the inflation rate, the inflation rate and the apparent time value of money must also be consistent. It was seen in Chapter 3 that the apparent time value of money is a measure more of inflation than of the real cost of investment capital. Inflation forecasts must at least be roughly consistent with the apparent time value of money. It is seldom correct to forecast inflation rates much larger than those currently existing while at the same time using an apparent time value of money that reflects money costs now. Most large firms establish a time value of money that is used for all economic calculations to ensure consistency within the firm. It would also be desirable to have a single projection of inflation for all economic calculations within the firm to further ensure consistency.

The details of price and inflation forecasting are beyond the scope of this text. Such forecasting is as much an art as a science. A review of past price forecasts will indicate that long-term price forecasts are not very reliable anyway and that the degree of sophistication in the forecast has little bearing on its accuracy. It is more important that the need for consistency among all parameters used in economic analyses be maintained. The most dangerous approach is to use several forecasts from different sources for different items. Each of these forecasts may be well thought out and be the best available for that particular item. However, these different forecasts may reflect different opinions and assumptions that strongly influence the final result. It is the author's opinion that for routine engineering economic analysis simple forecasts with the inflation and real price components clearly separated are generally quite satisfactory.

Forecasting requirements are less demanding if a real time value is specified and all transactions are stated in constant dollars. Only the real price changes need to be considered in this case. It is not normally necessary to forecast the inflation rate. There are two situations where it is necessary to estimate the inflation rate when using this approach, however. First, if time value is specified in apparent form, the inflation rate is required to convert the time value to real form. Also, if cash flows in the analysis are fixed in dollar terms (such as a contract would be), the inflation rate must be specified to convert the dollar cash flow to constant dollars.

It may seem at this point that the dollar approach and the constant dollar approach represent two basically different methods of making engineering economics calculations. However, closer scrutiny will show that under the assumption of constant inflation and constant time value the two approaches will give the same result. It was shown in Chapter 3 that equivalence calculations for dollar cash flows by use of the apparent time value are totally consistent with equivalence calculations made for the corresponding calculations by applying the real time value as long as inflation is constant and the relationship between real and apparent time values is specified by Equation 3.15. The relationship between dollar and constant dollar prices is identical to the relationship between dollar and constant dollar cash flows. As long as prices are projected as described here, any two transactions that are equivalent in dollar form will aslo be equivalent in constant dollar form. The important point to be made is that consistency must be maintained between the real and apparent time value, the inflation rate, dollar prices, and constant dollar prices.

4.2   THE CORRECT METHOD

There is no theoretical advantage of either the dollar approach or the constant dollar approach. It was just seen that once either the apparent time value or the real time value is specified, totally consistent results are obtained regardless of the approach used as long as all parameters are calculated consistently. The real theoretical problem comes when inflation, the real time value, and the apparent time value do not stay constant with time and consistency among these parameters breaks down. Discussion of this involved problem is postponed until Chapter 11. Selecting one of the methods at this point is largely a matter of personal preference and convenience.

The dollar approach has the advantage of expressing cash flows in the units in which the cash flows actually take place. Economic analysis often performs a planning function, stating what cash flows to expect when. Dollar representations of cash flows are more useful than constant dollar representations in this respect as they clearly show what money exchanges can be expected. Constant dollar representations must be adjusted for inflation to be used for this purpose.

The primary disadvantage of dollar representation is that it distorts one's "intuitive feel" for profitability assessment. Inflated cash flows in future years exaggerate future earnings and make marginal or unprofitable investments appear good. The apparent time value of money distorts the real value of capital and the real return that can be earned. Constant dollar representation avoids these problems.

The selection of one of the two approaches is not nearly as important as applying it correctly. Most large firms will have standardized their economic analysis procedures, so that this is not a choice of the engineer, anyway. Either approach will work. However, it is essential that care be used in applying the chosen method correctly. Above all, do not accidently mix the two methods. Be meticulous in determining whether a given cash flow is a dollar cash flow or a constant dollar cash flow.

4.3 YEAR END CONVENTION

It is customary, in the application of engineering economics, to group all cash flows that occur during a year and treat them as if they are a single cash flow occurring at the end of that year. This practice is referred to as the year end convention. The year end convention results in an economic distortion as is shown in Figure 4.1. The time value of money during the year is ignored, resulting in a cash flow somewhat less than the true economic value of the actual cash flow. The year end convention is justified on the basis of convenience and not on theoretical grounds.

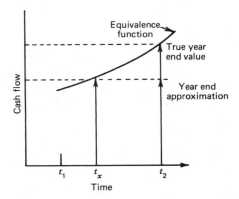

Figure 4.1 Distortion due to the year end convention.

Inflation can increase the distortion resulting from the year end convention and can cause some inconsistencies if not properly applied. Consider the purchase or sale of quantity Q of goods or services at time t_x as shown in Figure 4.2. Time t_x is between times t_1 and t_2, which are year ends. There are two ways to determine the year end cash flow. One is to calculate the cash flow according to the expected prices at time t_x and then treat that cash flow as if it occurred at time t_2. The second is to treat the purchase as if it occurred at time t_2 and calculate the year end cash flow by prices expected at time t_2. The latter approach causes less distortion when cash flows are expressed in dollars. The first approach also results in inconsistencies between constant dollar-real time value and dollar-apparent time value solutions.

The respective dollar and constant dollar cash flows for the transaction at time t_x are

$$Y^d(t_x) = Q \cdot p^d(t_x) \qquad (4.4)$$

and

$$Y^c(t_x) = Q \cdot p^c(t_x) \qquad (4.5)$$

These cash flows are considered to be year end cash flows at time t_2 using the first year end convention. The true equivalent values at time t_2 are

$$Y^d(t_2) = Y^d(t_x) \cdot [1 + i_a]^u \qquad (4.6)$$

and

$$Y^c(t_2) = Y^c(t_x) \cdot [1 + i_r]^u \qquad (4.7)$$

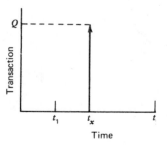

Figure 4.2   Transaction during year.

where

$$u = \frac{t_2 - t_x}{\Delta t}$$

The relative error due to the first form of the year end convention is simply the term

$$[1 + i_a]^u$$

when dollar notation is used and the term

$$[1 + i_r]^u$$

when constant dollar notation is used. Since i_a is normally larger than i_r, there is more distortion for the dollar notation than for the constant dollar notation. The result of this difference is that a slight discrepancy exists between dollar and constant dollar calculations.

The dollar cash flow by the year end convention can be converted to a constant dollar cash flow:

$$Y^{c'}(t) = \frac{1}{[1 + f](t_2 - t_0)/\Delta t} \cdot p^d(t_0) \cdot p^r \left(\frac{t_x}{t_0} \right) \cdot [1 + f]^{t_x - t_0/t}$$

where the ' is used to indicate that the constant dollar cash flow is determined from the dollar cash flow and the price is expressed in the form described previously in this chapter.

The expression for $Y^{c'}$ may be reduced to

$$Y^{c'}(t_2) = Q \cdot p^d(t_0) \cdot p^r \left(\frac{t_x}{t_0} \right) \cdot \left[\frac{1}{1+f} \right]^u \tag{4.8}$$

The constant dollar cash flow obtained directly by this year end convention can be expressed in a similar manner:

$$Y^c(t_2) = Q \cdot p^c(t_0) \cdot p^r \left(\frac{t_x}{t_0} \right) \tag{4.9}$$

Since

$$p^c(t_0) = p^d(t_0)$$

a comparison of Equations 4.8 and 4.9 shows that the discrepancy between the dollar cash flow and constant dollar cash flow using this year end convention is the factor

$$\left[\frac{1}{1+f}\right]^u$$

This discrepancy is not severe unless f is large since

$$0 < u < 1$$

However, readily apparent inconsistencies often occur when calculation results using dollar and constant dollar notation are compared.

The second form of the year end convention gives the following year end cash flows:

$$Y^d(t_2) = Q \cdot p^d(t_2) \tag{4.10}$$

and

$$Y^c(t_2) = Q \cdot p^c(t_2) \tag{4.11}$$

Equations 4.6 and 4.7 for the true equivalent year end values may be rewritten to give

$$Y^d(t_2) = Q \cdot p^d(t_x)[1 + i_a]^u \tag{4.12}$$

and

$$Y^c(t_2) = Q \cdot p^c(t_x)[1 + i_r]^u \tag{4.13}$$

Comparing Equations 4.10 and 4.11 obtained with the year end convention to the true values in Equations 4.12 and 4.13, it is seen that relative error due to this year end convention is

$$\frac{p^d(t_x)[1 + i_a]^u}{p^d(t_2)}$$

with the dollar notation and

$$\frac{p^c(t_x) \cdot [1 + i_r]^u}{p^c(t_2)}$$

for the constant dollar notation. If the price terms are expanded these factors
become

$$\frac{p^r(t_x/t_0)}{p^r(t_2/t_0)} \cdot \frac{[1 + i_a]^u}{[1 + f]^u}$$

and

$$\frac{p^r(t_x/t_0)}{p^r(t_2/t_0)} \cdot [1 + i_r]^u$$

respectively. The first factor readily reduces to the second one, and the
distortion is at least the same for both dollars and constant dollars using this
year end convention and consistent results will be obtained between dollar
and constant dollar calculations. When relative prices do not change rapidly

$$p^r\left(\frac{t_x}{t_0}\right) = p^r\left(\frac{t_2}{t_0}\right)$$

and the distortion here for both notations is the same as with the constant
dollar notation using the other form of the year end convention. These results
indicate that the second form of the year end convention should normally be
used unless there is reason to believe that major relative price changes will
occur between the time of actual transaction and the year end. That is, the
year end convention should be applied by treating the transactions as if they
occur at the year end and by using year end prices to evaluate cash flows.
However, the errors resulting if this rule is not followed are not too large
unless the inflation rate is fairly large and the dollar-apparent rate of return
method is being used. Also, it is important to use this form of year end
convention if it is desired to check solution results for consistency between
the two methods.

There are certain types of cash flows where this form of year end
convention is difficult to apply. Of particular importance is the situation
where cash flows are prescribed by contracts or other arrangements. The
concept of a year end price may have no real meaning in this case. The same
level of accuracy and consistency can be maintained by inflating the year end
cash flow as follows when dollars are being used

$$Y^d(t_2) = Y^d(t_x) \cdot [1 + f]^{[(t_2 - t_x)/\Delta t]} \tag{4.14}$$

If constant dollars are being used the specified dollar cash flows should be
converted to constant dollars at time t_x and then transposed to the year end:

$$Y^c(t_2) = Y^d(t_x) \cdot \frac{\bar{p}(t_0)}{\bar{p}(t_x)} \tag{4.15}$$

Other conventions for grouping cash flows are also used. One of the more common is the midyear convention where all cash flows for a given year are treated as if they occur in the middle of the year. The same principles that apply to the year end convention will also apply to these conventions.

4.4   EXAMPLES

Example 4.1

A heat pump manufacturer has an assembly line that produces 1500 heat pumps per year. The expansion valves for these heat pumps are purchased from another supplier. The company orders a supply of expansion valves every six months. An order was just placed and the valves cost $67 each. What cash flows can be expected for the expansion valves for the next three years? It is estimated that inflation will remain at approximately 12% annually over this period.

No price estimates for the expansion valves are given here. The only guidelines are the current price and the expected inflation rate. It is reasonable to assume that the relative price stays constant since no information to the contrary is available. The price then increases according to the inflation rate. The dollar price for each order is as follows:

Time from Present (years)	$\dfrac{\bar{p}(t)^a}{\bar{p}(0)}$	Price ($/valve)
0	1.00	67
½	1.058	71
1	1.120	75
1½	1.185	79
2	1.254	84
2½	1.328	89
3	1.405	94

$^a[1 + f]^{(t - 0)/1 \text{ yr}}$.

Each order should be for about 750 valves. The prices listed immediately above are required to determine the dollar cash flows. Since the relative price is assumed to stay constant, the price in constant dollars with today as the reference time should simply stay at $67 per valve. The cash flows for the next three years are then:

Time from Present (years)	Cash Flow	
	Dollars	Constant Dollars
½	53,250	50,250
1	56,250	50,250
1½	59,250	50,250
2	63,000	50,250
2½	66,750	50,250
3	70,500	50,250

Example 4.2

A particular large engine in an industrial plant is normally overhauled annually. A service company can be contracted to do this overhauling. They guarantee a price of $3800 for each overhaul for the next three overhauls. The engine has just been overhauled. This overhaul was typical and cost $1200 in parts and required 110 h of labor. The cost of the labor is $19/h. No information is available about expected price changes for engine parts, but the company expects labor costs to have about a 2% real increase each year. Inflation is forecast by this company to average about 10% for the next three years, and an apparent time value of money of 15% has been established. Determine the total equivalent value today of contracting to overhaul the engine and for the company making its own overhauls.

It is not stated whether these equivalent values should be in dollar terms or constant dollar terms, so the calculations are done both ways and first in dollar terms.

The dollar cash flows for the contract alternative are simply $3800 each overhaul since this price is guaranteed. The cost for the company to do the overhauls depends on the future prices of the parts and labor. No information is given about the future prices of parts, but it is reasonable to assume that their relative price stays the same. The relative price of labor is expected to increase 2% each year. The prices for each may be predicted by using Equation 4.3:

Time	Parts		Labor	
(years)	$p^r(t/t_0)$	price $/overhaul	$p^r(t/t_0)$	price ($/h)
0	1.00	1200	1.000	19.00
1	1.00	1320	1.020	21.32
2	1.00	1452	1.040	23.92
3	1.00	1597	1.061	26.84

The dollar cash flows can be determined from these prices and are as follows:

Time (years)	1	2	3
Overhaul cost ($)	3665	4083	4549

The equivalent value of each cash flow can be calculated by the 15% apparent time value:

Time (years)	Contract		Company	
	Cash Flow ($)	Equivalent Value at Time 0 ($)	Cash Flow ($)	Equivalent Value at Time 0 ($)
1	3800	3304	3665	3187
2	3800	2873	4083	3087
3	3800	2499	4549	2991
Total		$8676		$9265

It would appear that the total equivalent value for the contract is less than the total equivalent cost for the company to do the overhauls. However, this result depends on a number of assumptions: the future inflation rate, prices, and whether the service company will actually perform as promised.

This problem can also be solved using the constant dollar-real time value approach. The constant dollar prices, using today as the reference time, can be calculated by direct application of Equation 4.2. The relative prices were calculated previously. The results are as follows:

Time (years)	Parts ($/overhaul)	Labor ($/h)
0	1200	19.00
1	1200	19.38
2	1200	19.76
3	1200	20.16

The resulting constant dollar cash flows are:

Time (years)	1	2	3
Overhaul cost (%)	3332	3374	3418

The dollar cost of having the service company do the overhaul remains constant. The constant dollar cost may be found by application of Equations 2.9 and 2.10. The resulting constant dollar cash flows are:

Time (years)	1	2	3
Overhaul cost ($)	3456	3140	2855

The equivalent constant dollar cost for each set of cash flows can now be found by using the real time value. The real time value must first be found:

$$i_r = \frac{1.0 + 0.15}{1.0 + 0.10} - 1.0 = 0.0455 = 4.55\%$$

The resulting equivalent values can now be calculated:

Time (years)	Contract Cash Flow ($)	Contract Equivalent Value at Time 0 ($)	Company Cash Flow ($)	Company Equivalent Value at Time 0 ($)
1	3456	3304	3332	3187
2	3140	2873	3374	3087
3	2855	2499	3418	2991
Total		$8676		$9265

The results are exactly the same as before. However, these total equivalent cash flows are in constant dollars, and the previous results were in dollars. The numerical results would not be identical except that the reference time for the constant dollars is the same as for the equivalent value, and therefore dollars and constant dollars are the same. If a different time had been chosen for either the equivalent value or the reference time, the results would not be numerically the same, but the same relative difference would exist between the equivalent cash flows for the two alternatives.

Example 4.3

An oil company that operates an oil field in a very severe cold climate is considering a project to build enclosures for each well pumping unit to facilitate inspection and servicing during the winter. The enclosures would be built over a five year period. The four main cost items in constructing the enclosures are labor, lumber, galvanized sheet metal, and concrete. The following economic information has been supplied. How should one proceed with a cost analysis?

Cost of lumber at field	$0.60 per board foot
Cost of money	12% annually

	Cost at Field		
Time from Present (years)	Labor ($/hr)	Sheet Metal ($/100 ft²)	Concrete ($/yd³)
0	12.50	50.00	90
1	13.00	57.50	105
2	13.75	65.00	120
3	14.50	75.00	135
4	15.25	90.00	160
5	16.00	100.00	180

It would be proper to check these price forecasts for consistency. It is immediately seen that the cost of lumber is shown as staying constant, or at least no information is given about an increase, whereas the other cost components are shown to increase substantially. There should be some justification for this difference before any economic analysis is made. It may also be interesting to examine the price increase for each of the other three items:

	Percent Price Increase Over Previous Year		
Year	Labor	Sheet Metal	Concrete
1	4.0	15.0	16.7
2	5.8	13.0	14.3
3	5.5	15.4	12.5
4	5.2	20.0	18.5
5	4.9	11.1	12.5
Average	5.1	14.9	14.9

The price forecasts for the sheet metal and the concrete appear reasonably consistent on a per annum basis and are very consistent on the average over the five year time period. They would appear to reflect an inflation rate of about 15%. The variations from year to year could be accounted for by market fluctuations. The cost of labor is shown as increasing much less rapidly. Again, this discrepancy should be explained. One further item that should be examined is the time value of money. Presumably the 12% figure is an apparent time value. This value is inconsistent with the approximately 15% inflation shown for the sheet metal and concrete. It is unlikely that a company will have an apparent time value less than the inflation rate for five years. The 12% time value might be reasonable for the 5% inflation indicated by the labor cost but not the 0% inflation indicated by the lumber cost.

At this point the engineer should not proceed with an economic analysis until these discrepancies are explained and corrected if necessary. It is entirely possible, but unlikely, that all this economic information was based on the same general set of assumptions. Unless the consistency can be verified, an economic analysis based on this information could be very misleading. This example demonstrates the importance of first making assumptions about future inflation and then estimating future prices and the time value of money.

Example 4.4

A steam generator in a particular manufacturing plant runs about 90% of the time, being shut down only for maintenance and repairs or when the plant must be shut down. The steam generator consumes 10,000 ft³ of natural gas per hour when in operation. The company is billed at the end of each month for that month's gas consumption. The present price of the natural gas is $5.20/100 ft³ and is expected to increase about 1% each month. This price increase is slightly greater than the expected rate of inflation, which is 11% annually. This difference corresponds to about a 1.5% annual relative price increase for the natural gas. The difference is due to anticipated decline in natural gas availability. The company has established a time value of money of 15%, which is reasonable considering the expected 11% inflation. What is the equivalent cost now of the next four years' fuel? If constant dollars are used, the company prefers that you work in 1967 dollars and use the GNP-IPD as a measure of inflation. The present value of the GNP-IPD is 250, using 1967 as the reference year.

The gas is billed monthly, and there will be a cash flow each month. However, it is more reasonable to use the year end convention and treat each year's gas purchases as a single cash flow. The proper, and in this case the simplest, way to generate the year end cash flow is to determine the annual fuel consumption and treat it as if all the gas is to be purchased at the year end price.

The annual gas consumption is

$$8760 \ \frac{h}{yr} \cdot 0.90 \cdot 10,000 \ \frac{ft^3}{h} = 78.8 \times 10^6 \ \frac{ft^3}{yr}$$

The dollar price of the gas is expected to increase 1% each month. The increase for one year would then be

$$1.01^{12} - 1.0 = 0.127 = 12.7\%$$

The dollar prices and the resulting year end cash flows are then as follows:

Year	Price ($/10³ ft³)	Cash Flow ($ × 10³)
1	5.86	461.7
2	6.60	520.3
3	7.44	586.3
4	8.38	660.3

The equivalent value at time 0 can be calculated for these year end cash flows using the 15% apparent time value of money:

Year of Cash Flow	Equivalent Value at Time 0 ($ × 10³)
1	401.5
2	393.4
3	385.5
4	377.7
Total	1558.1

The next four years' fuel cost is equivalent to $1,558,000 today.

The calculation could also be made in constant dollar terms. The present price in 1967 dollars is

$$[\$5.20/1000 \text{ ft}^3] \cdot \frac{\$100}{\$250} = \$2.08/1000 \text{ ft}^3$$

The relative price increase each year is

$$\frac{1.01^{12}}{1.11} - 1.0 = 0.0152 = 1.52\%$$

The constant dollar price and the constant dollar cash flows can now be readily calculated:

Year	Price ($/1000 ft³)	Cash Flow ($ × 10³)
1	2.11	166.4
2	2.14	168.9
3	2.18	171.5
4	2.21	174.1

The real time value must be found so that the equivalent values of these cash flows can be calculated. The real time value is

$$\frac{1.15}{1.11} - 1.0 = 0.036 = 3.6\%$$

The equivalent value of the constant dollar cash flows can now be calculated.

Year of Cash Flow	Equivalent Value at Time 0 ($\bar{\$} \times 10^3$)
1	160.6
2	157.4
3	154.2
4	151.1
Total	623.3

The next four years' of fuel expense has an equivalent cost today in 1967 dollars of $623,300.

We can verify that the two solutions are equivalent by converting this quantity to today's dollar:

$$\bar{\$}623.3 \times 10^3 \frac{\$250}{\bar{\$}100} = \$1558 \times 10^3$$

This result is consistent with the earlier calculation.

CHAPTER FIVE

Present Worth

It is assumed in this and the next two chapters that the reader has some familiarity with basic engineering economic calculations. Chapter 4 described how inflation should be introduced into engineering economy calculations. Chapters 5, 6, and 7 apply this theory to the three most common methods of comparing alternatives in engineering economics; present worth, annual cash flow, and internal rate of return. The theory behind these methods is briefly reviewed; however, a reader who is not familiar with these calculations may wish to review them in any good basic engineering economics text. It is also assumed that the reader is familiar with interest factors and interest tables. Equations for interest factors are given in Appendix C.

5.1 PRESENT AND FUTURE VALUE

The purpose of engineering economics is to allow the selection of one of two or more engineering alternatives by rational economic criteria. The principle of equivalence allows any group of cash flows to be reduced to a single equivalent cash flow. Thus the incomes and expenses for a given alternative may be reduced to a single equivalent cash flow. This equivalent cash flow can be calculated for any point in time. The net equivalent cash flows for two or more alternatives may be compared as shown in Figure 5.1. The curves show the net equivalent value of each alternative as a function of time. For a single time value, these curves will never intersect. It is stated here, without further proof, that with everything else being equal, the alternative with the higher curve is preferred.

All that is necessary to determine which alternative has the higher curve is to evaluate the equivalent values for the same point in time. The alternative with the larger equivalent value at the common point in time will have the larger equivalent value at any other point in time and, thus, is the preferred

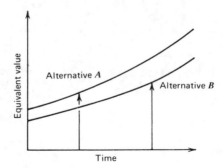

Figure 5.1   Comparing the equivalent value of two alternatives.

alternative. This common time is normally taken as the present time and hence terms such as "present value" and "present worth" are applied. However, any other time may be selected for evaluating the equivalent values of the alternatives as long as it is the same for all alternatives. The following discussion refers only to present value calculations. It should be remembered that the comments pertain to any common time comparison of equivalent values.

Inflation requires that an additional consideration be included before the present value comparison be made. The equivalent values must be expressed in dollars of the same value. If the equivalence calculations are made by using dollar statements of individual cash flows and the apparent time value, the resultant equivalent value is expressed in dollars. The value of these dollars is determined by the time at which the equivalent value is taken. That is, it is expressed in dollars of the time of the present value. If the equivalence calculations are made by using constant dollar statements of the individual cash flows and the real time value, then the resultant equivalent value is expressed in the same constant dollars.

It is usually best to work strictly in dollars or strictly in constant dollars to avoid confusion. However, it is not uncommon to have engineering economics problems where it is convenient to evaluate one alternative by using dollars and the apparent time value and to evaluate another alternative by using constant dollars and real time value. A comparison of the results of such a calculation is straightforward. Equivalent cash flows may be treated just like any other cash flows and thus may be expressed in any constant dollar. Recall that

$$Y^c = Y^d(t) \ \frac{\bar{p}(t_0)}{\bar{p}(t)} \tag{5.1}$$

where Y^c is the constant dollar expression of the dollar cash flow Y^d that occurs at time t and t_0 is the reference time for the constant dollar. The equivalent cash flow expressed in dollars may be converted to a constant dollar equivalent cash flow and vice versa by applying this relationship. If t_0 is taken to be the present time and the equivalent values are calculated for the present time, the ratio

$$\frac{\bar{p}(t_0)}{\bar{p}(t)}$$

is unity. Dollar equivalent values and constant dollar equivalent values may be compared directly with each other in this case. This result makes it easy to compare equivalent values for different alternatives since it is usually easiest to do constant dollar calculations in today's dollar (t_0 = today). However, you should never forget that this situation is a special case, and in general Equation 5.1 must be used to convert between dollar and constant dollar present values.

5.2 CAPITALIZED COST

Capitalized cost calculations are merely a specialized case of the present value calculation. It is the case where the economic analysis period extends over an infinite period of time. Such problems are solvable only if it can be shown that the cash flows over the infinite period of time result in a finite equivalent value. When the cash flows repeat periodically, it is easily shown that the resulting infinite series of cash flows has a finite equivalent value. The present value of such a series can be calculated with the series present worth factor:

$$P = A \cdot \left(\frac{P}{A} \, i, \, n \right)$$

where P is the present value, A is the magnitude of periodic cash flow, and $(P/A, i, n)$ is the series present worth factor. The series present worth factor may be expressed mathematically to yield

$$P = A \left\{ \frac{1}{i} - \frac{1}{i \cdot [1 + i]^n} \right\} \qquad (5.2)$$

where i is the appropriate time value and n is the number of time periods.

Note that Δt, the compounding period, will be one year here only if the cash flows repeat every year; otherwise, it may be something different. For an infinite time period, n becomes infinity and Equation 5.2 reduces to

$$P = A \cdot \frac{1}{i} \qquad (5.3)$$

provided that $i > 0$.

Inflation results in the dollar cash flows increasing with time for most engineering alternatives. Even if the same operation is performed year after year, the same purchases are made and the same sales made, the dollar cash flows will increase each year as a result of inflation, and it is impossible to calculate the equivalent cash flow by applying the series present worth factor. The constant dollar cash flow will repeat periodically if the same purchases and sales are made and the relative prices do not change. It is often reasonable to assume that relative prices do stay constant; thus it is usually most convenient to work in constant dollar cash flows and the real time value when doing capitalized cost calculations.

It is not essential that all cash flows repeat periodically to solve a capitalized cost calculation. The problem is somewhat more complicated in other situations, however. The equivalent value of a set of cash flows may be calculated in general by

$$Y(t) = \sum_{j=1}^{\infty} Y_j \cdot \left[\frac{1}{1+i} \right]^{[(t_j - t)/\Delta t]} \qquad (5.4)$$

where $Y(t)$ is the equivalent value at time t, Y_j is an individual cash flow at time t_j, and i is the appropriate time value for time period Δt.

A special case of Equation 5.4 can be solved when Y_j changes by the same percentage each time period; thus

$$Y_j = Y_0 \cdot [1 + s]^j \qquad (5.5)$$

where s is the fractional increase each time period. The equivalent value of such a series of cash flows is

$$Y(t) = \sum_{j=1}^{\infty} Y_0 \cdot [1 + s]^j \cdot \left[\frac{1}{1+i} \right]^{[(t_0 + j \cdot \Delta t - t)/\Delta t]} \qquad (5.6)$$

where t_0 is one time period Δt before the first cash flow. Equation 5.6 can be rearranged to

$$Y(t) = \sum_{j=1}^{\infty} Y_0 \cdot [1 + s]^j \cdot \left[\frac{1}{1 + i}\right]^j \cdot [1 + i]^{[(t - t_0)/\Delta t]}$$

or

$$Y(t) = Y_0 \cdot [1 + i]^{[(t - t_0)/\Delta t]} \cdot \sum_{j=1}^{\infty} \left[\frac{1 + s}{1 + i}\right]^j \qquad (5.7)$$

The summation can be simplified since

$$\frac{1 + s}{i - s} = \sum_{j=1}^{\infty} \left[\frac{1 + s}{1 + i}\right]^j \qquad (5.8)$$

as long as $s > -1$, $i > -1$, and $i > s$.*
Equation 5.7 then reduces to

$$Y(t) = Y_0 \cdot [1 + i]^{[(t - t_0)/\Delta t]} \cdot \frac{1 + s}{i - s} \qquad (5.9)$$

If we further let $t = t_0$, which is analogous to the timing of the cash flows used for the series present worth factor, Equation 5.9 reduces to

$$Y(t) = Y_0 \cdot \frac{1 + s}{i - s} \qquad (5.10)$$

This equivalence relationship is shown in Figure 5.2.

This derivation shows that when cash flows increase periodically as described by Equation 5.5, a finite equivalent amount can be found for a capitalized cost calculation as long as $s < i$. If $s > i$, there is no finite solution. Dollar cash flow increasing due to a constant inflation rate or due to a constant inflation rate and a constant percentage increase in relative price behave as described by Equation 5.5. Constant dollar cash flows increasing due to a constant percentage increase in a relative price behave this way also. Thus Equation 5.10 may prove useful for calculating equivalent cash flows in many capitalized cost problems.

One question that has no meaning in the mathematical world but must be answered in engineering economics is, "How long is an infinite time period?"

Summation of Series, L. B. W. Jolley, 2nd Rev. Ed., Dover Publ., New York, 1961, p. 8, Series No. 39.

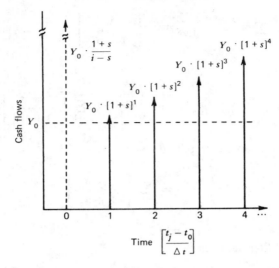

Figure 5.2 Equivalent value of an infinite series of exponentially increasing cash flows.

No engineering economics analyses can be considered for a truly infinite time. Few analyses have any relevance for even a century. At what point then does an engineering economics problem become a capitalized cost problem? There is no clear-cut distinction between capitalized cost situations and finite life situations. As the time period becomes longer, the results become increasingly like those of the capitalized cost situation. At some point the results are the same for all practical purposes.

One method that is sometimes used to estimate how long an analysis must extend before it should be treated as a capitalized cost calculation is to use the series present worth as a guide. A finite series of uniform cash flows is no different economically from an infinite series of the same cash flows when the series present worth factor for the finite series is approximately equal to the value for an infinite length of time. Thus it is said that the length of time required for the series present worth factor to become approximately equal to the value for an infinite length of time is a good guide as to when an analysis should be considered a capitalized cost situation. Such a decision is roughly equivalent to saying that the cash flows after the length of time given by this criterion do not really matter in the economic analysis.

This criterion is very dangerous to apply when dealing with apparent time values. The reason for this danger is that the criterion requires the cash flows to be the same period after period. The apparent time value implies that cash flows are stated in dollars and will almost certainly increase from time period to time period if there is any significant inflation. It is much more reasonable to apply this criterion when the real time value is being used. The constant dollar cash flows are likely to be nearly the same from time period to time period at least over a long time span.

The consequence of this argument is that much longer time periods may be needed before an analysis becomes a capitalized problem than many people may think, and cash flows over a much longer period of time are important for a valid economic analysis. Example 5.3 shows that analysis periods from 75 to 100 years may be required before a capitalized cost calculation should be made. This result also implies that cash flows cannot be ignored 75 to 100 years from now simply because of the diminishing effect on their importance due to the time value of money. However, other factors such as technical obsolescence or need for a quick payback of the invested capital to maintain liquidity may establish upper limits on the length of the analysis in many cases.

5.3 EXAMPLES

Example 5.1

A manufacturing plant has an outdated boiler that produces 200 psi steam at 500°F. The boiler is not very efficient and presently uses about 3000 BTU of fuel per pound of steam. The plant requires about 10^6 lb of steam each month. Fuel for the boiler presently costs \$2.10 per 10^6 BTU. It is planned to replace the boiler in five years as part of a plant modernization. To reduce fuel use in the meantime, a modification costing \$37,000 has been proposed. This modification should reduce fuel consumption to 2600 BTU per pound of steam.

The company has established a time value of money of 19% and has projected the inflation rate to be about 11%. The relative price of the fuel is expected to increase 5% each of the next two years and then should stay about constant. Given these conditions, is the modification worthwhile?

Two alternatives exist here, the alternative of doing nothing and leaving the boiler as it is and the alternative of modifying the boiler. The preferable alternative may be found by calculating the net savings of the modification and weighing these against the expense. It is clear here that an analysis period of five years is appropriate as the boiler will be eliminated then.

The annual fuel savings is

$$\left[3000 \, \frac{\text{BTU}}{\text{lb}} - 2600 \, \frac{\text{BTU}}{\text{lb}} \right] \cdot 10^6 \, \frac{\text{lb}}{\text{month}} \cdot 12 \, \frac{\text{month}}{\text{yr}} = 4.8 \times 10^9 \, \text{BTU}$$

The economics can be evaluated in either dollar terms or constant dollar terms. Let us do it in constant dollar terms first. The relative prices are shown in the following list along with the constant dollar prices and constant dollar cash flows, with the reference time set at the present ($t_0 = 0$):

Year	Relative Price	Price of Fuel ($/10^6 BTU)	Savings ($)
1	1.050	2.205	10,584
2	1.1025	2.315	11,113
3	1.1025	2.315	11,113
4	1.1025	2.315	11,113
5	1.1025	2.315	11,113

The present value of these cash flows and the initial expense can be found, but first the real time value must be found. Presumably, the 19% figure is an apparent time value:

$$i_r = \frac{1.19}{1.11} - 1.0 = 0.0721 = 7.21\%$$

The net present value of the modification in constant dollar terms can now be found. The $37,000 expense will be the same in constant dollars as in dollars since it will occur at approximately the present time and the reference time is the present:

$$PV = -\$37,000 + \left(\frac{P}{F}, 7.21\%, 1\right) \cdot [\$10,584$$
$$+ \left(\frac{P}{A}, 7.21\%, 4\right) \cdot \$11,113]$$
$$= \$7820$$

The modification has a net positive present value of $7820 in today's dollars.
    The present value could equally well have been calculated by using dollar statement of the cash flows. The prices and cash flows are calculated much the same as before except the price increase due to inflation must be included:

Year	Relative Price	$\frac{\bar{p}(t)}{\bar{p}(t_0)}$	Price of Fuel ($/10^6$ BTU)	Savings ($)
1	1.050	1.110	2.448	11,748
2	1.1025	1.232	2.853	13,693
3	1.1025	1.368	3.166	15,199
4	1.1025	1.518	3.515	16,871
5	1.1025	1.685	3.901	18,726

The present value of each cash flow can now be found by using the 19% apparent time value:

$$PV = -\$37,000 + \left(\frac{P}{F}, 19\%, 1\right) \cdot \$11,748 + \left(\frac{P}{F}, 19\%, 2\right) \cdot \$13,693$$
$$+ \left(\frac{P}{F}, 19\%, 3\right) \cdot \$15,199 + \left(\frac{P}{F}, 19\%, 4\right) \cdot \$16,871$$
$$+ \left(\frac{P}{F}, 19\%, 5\right) \cdot \$18,726$$
$$= \$7820$$

The result is identical to the previous calculation since the constant dollar reference time and the time of the present value are the same.

Example 5.2

An electronics company is purchasing a new computer system for one of its sales offices. As part of the purchase, the company has the option to enter into a four year maintenance contract. The computer supplier will provide all servicing for $2850 per year with the $2850 to be paid at the beginning of each year. The alternative to this contract is to have the computer supplier service the computer as needed. Experience with similar computer systems in the company shows that an average of 1.1 service calls per month are required. The present cost of a service call is $100 per call plus $22 per hour of servicing. The average service call requires 4.9 h. The computer system comes with a warranty that covers all parts required in the servicing.

The electronics company has an apparent time value of 16% and projects inflation to be 10% for the next several years. Should the service contract be used?

In this analysis it is convenient to assess one alternative in dollar terms and the other alternative in constant dollar terms. The cost of the contract stays constant in dollar terms. It is a very simple matter to calculate its present cost:

$$PC = \$2850 + \left(\frac{P}{A}, 16\%, 3\right) \cdot \$2850$$

$$= \$9250$$

The "service as needed" option has its costs stated at today's rates. There is no reason to believe they will not increase approximately at the same rate as inflation since no contrary information is supplied. Therefore, these costs will stay constant in constant dollar terms. The annual cost in constant dollars with today as the reference time is

$$12\ \frac{\text{month}}{\text{yr}} \cdot 1.1\ \frac{\text{call}}{\text{month}} \cdot [\$100/\text{call} + \$22/\text{h} \cdot 4.9\ \text{h/call}] = \$2740/\text{yr}$$

The real time value is

$$\frac{1.16}{1.10} - 1.0 = 0.0545 = 5.45\%$$

The present cost of this option can now be calculated:

$$PC = \left(\frac{P}{A}, 5.45\%, 4\right) \cdot \$2740$$

$$= \$9620$$

These present costs can be compared directly since they both occur at the same time and this time is also the reference time for the constant dollar. The contract appears to have a slightly lower net present cost and thus is probably the preferred option.

It might be interesting to note what our result would be if we were ignorant of inflation. We most likely would have assumed that the present charges for servicing would stay the same (in dollars) and would have evaluated the second alternative as follows:

$$PC = \left(\frac{P}{A}, 16\%, 4\right) \cdot \$2740$$

$$= \$7670$$

This incorrect calculation would show the "service as needed" option to be substantially less costly.

Example 5.3

A realistic real time value is about 4% annually for many companies. Recent inflation rates would translate this quantity into an apparent time value of about 15%. Using the series present worth criterion discussed before, determine how long an analysis period is required before a calculation should be considered a capitalized cost calculation.

The proper interest rate to use for this decision is the real time value. The value of the series present worth factor is shown as a function of the analysis period in Figure 5.3. It is seen that it takes 80–100 years for the value of the factor to approach its ultimate value. Thus analysis periods shorter than this should not be considered capitalized cost calculations with this real time

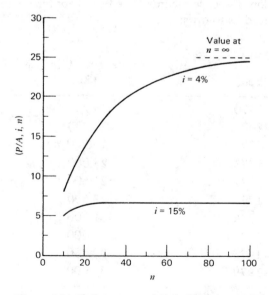

Figure 5.3 Series present worth for long time periods.

value. The value of the series present worth using the 15% apparent time value is also shown in Figure 5.3. It is seen how misleading the use of the apparent time value can be. The factor reaches its ultimate value within only 25–30 years with the 15% interest.

These results tell us that expenses or incomes that repeat periodically are important for a very long period of time if we assume that they increase along with prices in general (stay constant in constant dollar terms). The series present worth factor with the apparent time value is used in a capitalized cost problem only if dollar cash flows repeat, which is seldom the case. The results in Figure 5.3 also show that long-term expense or incomes are much more important than may be suspected. The long-term value of the series present worth factor is 25 in this problem, meaning that a continuing expense is equivalent to approximately 25 times its present annual amount. If we mistakenly make this evaluation using the apparent time value, we would say that a continuing expense is only equivalent to about 6.7 times its annual amount.

Example 5.4

The water for a particular irrigation system is brought in from a stream through a dirt canal. This canal requires extensive maintenance every year because of errosion. Expense records have been kept for a number of years and are summarized as follows:

Year	Maintenance Expense ($)
1970	16,250
1971	22,700
1972	14,900
1973	28,000
1974	21,600
1975	23,500
1976	31,200
1977	25,600
1978	32,700
1979	34,000
Average annual cost	25,045

The canal could be lined with concrete effectively eliminating this expense. Similar canals lined over 50 years ago are still in very good condition. A cost estimate shows that it would cost about $400,000 to line the canal now. The owners have established a time value of money of 17% and expect inflation to be about 12%. Is it economical to line the canal? The irrigation system will be used for as long as one can forsee into the future. The current value of the GNP-IPD is about 250.

The first thing to realize here is that the average annual maintenance expense calculated for us has no meaning. The value of a dollar decreased by a factor of about 2 from 1970 to 1979. The most meaningful way to assess these costs is to express them all in constant dollars. The expenses are restated here in constant dollars using 1967 as a reference time:

Year	Maintenance Expense ($)	Maintenance Expense (1967 $)
1970	16,250	14,040
1971	22,700	18,680
1972	14,900	11,770
1973	28,000	20,910
1974	21,600	14,710
1975	23,500	14,600
1976	31,200	18,440
1977	25,600	14,270
1978	32,700	17,000
1979	34,000	16,220
Average		16,060

This average cost could be expressed in today's dollar if desired, but it is just as easy to leave it in 1967 dollars. There is no noticable trend in the constant dollar costs, and since no other information is available, it seems reasonable to assume that the $\overline{\$}16,060$ average would be typical of the future. The life of the lining appears to be quite long, possibly 75–100 years or more. Also, the canal is likely to be used for this long so a capitalized cost analysis is appropriate. The $400,000 construction expense needs to be stated in 1967 dollars:

$$\$400,000 \cdot \frac{100\$}{250\$} = \$160,000$$

The real time value of money must also be found:

$$i_r = \frac{1.17}{1.12} - 1.0 = 0.0446 = 4.46\%$$

The net present value of the lining is then

$$PV = -\$160,000 + \left(\frac{P}{A}, 4.46\%, \infty\right) \cdot \$16,060$$

$$= \$199,700$$

It is not necessary, but the present value may be expressed in today's dollars:

$$PV = \$199,700 \cdot \frac{250\$}{100\$}$$

$$= \$499,400$$

Either way, the lining appears to be a worthwhile investment.

It is interesting to determine the result that would be obtained if inflation were ignored. The average annual savings would be $25,045, the investment cost $400,000, and the interest 17%. The net present value would then be

$$PV = -\$400,000 + \left(\frac{P}{A}, 17\%, \infty\right) \cdot \$25,045$$

$$= -\$252,700$$

This erroneous value shows the lining to be a poor investment and points out the importance of properly including inflation.

CHAPTER SIX

Annual Cash Flow

It was shown in Chapter 5 that comparison of equivalent values of various alternatives was the proper theoretical means of making an economic choice from among these alternatives as long as the equivalent values were calculated for the same point in time and are expressed in dollars of the same value. These criteria may be expanded to include other equally correct methods of comparison. One of the most common methods is the annual cash flow calculation. A present value can be converted to an economically equivalent uniform annual cash flow by the relationship

$$ACF = PV \cdot \left(\frac{A}{P}, i, n \right) \qquad (6.1)$$

where *ACF is the uniform annual cash flow and PV* is the present value. This process is sometimes referred to as *annualizing* or *annualization*. Of course, the annual cash flow may be calculated directly from an alternative's cash flows rather than first calculating the present value if that is more convenient. It is easily shown that the annual cash flow is a proper variable for comparison. Consider two alternatives such that

$$ACF_1 > ACF_2 \qquad (6.2)$$

Equation 6.1 may be used to substitute for the annual cash flow in this inequality:

$$PV_1 \cdot \left(\frac{A}{P}, i, n \right) > PV_2 \cdot \left(\frac{A}{P}, i, n \right) \qquad (6.3)$$

As long as both alternatives are annualized over the same time period n, the interest factor drops out of the inequality and it reduces to

$$PV_1 > PV_2 \qquad (6.4)$$

Thus the alternative with the greatest annual cash flow will also have the greatest present value provided that both alternatives are annualized over the identical time period. This last restriction proves to be an important limitation later on. The length of time for the annualization is arbitrary as long as it is the same for both alternatives. The obvious choice, however, is the length of the analysis period.

6.1 DOLLAR AND CONSTANT DOLLAR ANNUALIZATION

The preceding justification for the annual cash flow method did not specify whether the cash flows should be expressed in dollars or constant dollars. The relationships expressed are equally valid if the dollar-apparent time value or the constant dollar-real time value approach is used. It is very important to avoid confusion between the two methods, however. With the present worth method, the equivalent present values of two alternatives could be compared even if one was achieved with the dollar-apparent time value method and the other by the constant dollar-real time value method as long as the reference time for the constant dollar was at time "zero." Such a situation does not exist for the annual cash flow method. Annual cash flows calculated by the dollar-apparent time value method cannot be compared to annual cash flows calculated by the constant dollar-real time value method. Even though alternatives may be evaluated and compared using either form of the annual cash flow, the two forms are quite different and have different meanings.

An annual flow expressed in constant dollars represents the same flow of value occurring each year. It is roughly equivalent to having the same amount of goods and services exchanged each year. The value of each year's cash flow is exactly the same as the others. Year by year equivalent dollar cash flows will increase each year, as shown in Figure 6.1. Similarly, an annual cash flow expressed in dollar terms does not represent a constant exchange of goods and services from year to year. The value of each cash flow is different, only the dollar amount stays the same. The year by year equivalent constant dollar cash flows decrease each year as shown in Figure 6.2. Thus a constant dollar annual cash flow is something entirely different from a dollar annual cash flow.

It is possible to find constant dollar annual cash flows and dollar annual cash flows that are economically equivalent to each other. Equation 6.1 may be stated in either dollar or constant dollar terms:

$$ACF^d = PV^d \cdot \left(\frac{A}{P}, i_a, n \right) \tag{6.5}$$

$$ACF^c = PV^c \cdot \left(\frac{A}{P}, i_r, n \right) \tag{6.6}$$

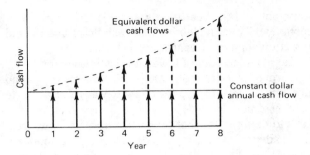

Figure 6.1 Constant dollar annual cash flow.

where the d superscript denotes dollars and the c superscript denotes constant dollars. The relationship between the respective present values is

$$PV^d = PV^c \; \frac{\bar{p}(t)}{\bar{p}(t_0)} \tag{6.7}$$

where t is the time of the present value and t_0 is the reference time for the constant dollar. The relationship between the two annual cash flows is then

$$\frac{ACF^d}{ACF^c} = \frac{\bar{p}(t)}{\bar{p}(t_0)} \cdot \frac{(A/P, i_a, n)}{(A/P, i_r, n)} \tag{6.8}$$

If the reference time and the present value time are the same ($t = t_0$), the ratio of the annual cash flows is simply the ratio of the interest factor $(A/P, i, n)$

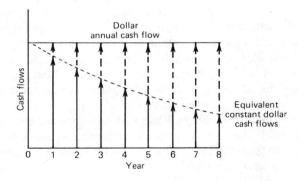

Figure 6.2 Dollar annual cash flow.

using the apparent and the real time values. It is important to realize that these are economically equivalent annual cash flows and the individual cash flows are not equivalent on a year by year basis.

Normally, it is best to solve for annual cash flows for all alternatives in only dollar terms or only constant dollar terms to avoid confusion. Occasionally, it is easier to solve for one alternative in dollar terms and the other in constant dollar terms. Equation 6.8 may then be used to convert from one form to the other so that the results may be compared. There is no real theoretical advantage of using the dollar-apparent time value method or the constant dollar-real time value method for annual cash flow comparison. When applied correctly, both give the same result. The constant dollar representation may have some conceptual advantage since it does represent a true uniform annual amount in value terms. The dollar annual cash flow represents a different quantity each year since the units of value per dollar change each year. It should be realized, however, that both forms are only equivalent representations of actual cash flows and do not describe the year by year cash flows.

6.2   UNEQUAL LIFE ALTERNATIVES

One of the most useful applications of the annual cash flow method for economic analysis is for the comparison of alternatives with unequal lives. Often, two alternatives that perform the same function but have different service lives must be compared. A common method of making this comparison is to evaluate the equivalent annual cost of both alternatives and make a choice based on these costs. There are some limitations on this method, however, particularly when inflation is present.

The validity of the annual cash flow comparison comes from its equivalence to a present value comparison, as has already been shown. However, the annual cash flow comparison is the same as a present value comparison only when the annualization is performed for all alternatives over the same time period n in Equation 6.1. A present value comparison of unequal life alternatives requires that an analysis time period be established such that both alternatives serve an integer number of service lives. This requirement ensures that both alternatives are treated fairly in that the analysis period does not require one alternative to be terminated in the middle of its service life. An annual cash flow comparison could be made over this same extended analysis period. If the cash flows associated with an alternative repeat from one life cycle to the next, the annual cash flow for each life cycle will be the same and the annual cash flow over the entire analysis period will be the same as the annual cash flow for a single life cycle, as is shown in Figure 6.3.

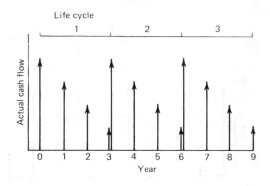

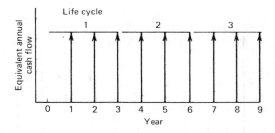

Figure 6.3   Annual cash flow for succeeding life cycles where cash flows recur.

The comparison of single, unequal life, annual cash flows is equivalent to the present value comparison only if this condition is met. If the cash flows do not repeat from one life cycle to the next, the annual cash flow for each life cycle is different and there is not a direct relationship between the annual cash flow for a single life cycle and the annual cash flow for the entire analysis period. This later situation is shown in Figure 6.4.

Unfortunately, inflation almost always results in dollar cash flows that increase from one life cycle to the next. Thus it is seldom correct to compare unequal life alternatives by using the annual cash flow for single life cycles with dollar representation of cash flows where lives of the alternatives are different. Even when the cash flows during the life cycle reflect price increases resulting from inflation, this method should not be used with dollar representations. There is a good chance that constant dollar cash flow will recur from one life cycle to the next, at least approximately so. The only requirement for this to happen is that relative prices remain constant. Since this assumption should normally be made when no contrary information exists, it is usually possible to use this technique to compare alternatives with

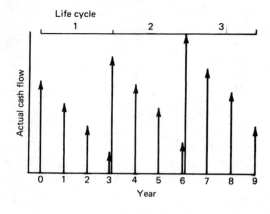

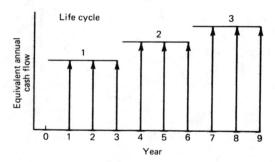

Figure 6.4   Annual cash flow for succeeding life cycles where cash flows do not recur.

unequal lives as long as constant dollar representation of the cash flows is used.

The existence of nonconstant relative prices implies that the annual cash flow of single life cycles should not be used for comparison of unequal life alternatives regardless of whether dollar representation or constant dollar representation of the cash flows should be used. In fact, the present value solution with integer numbers of life cycles may also be invalid if the relative prices change greatly. One alternative may be preferable for some time period, followed by a switch to the other alternative. Such analyses are much more complex than those discussed here, and data accuracy is seldom sufficient to justify the added level of detail in the calculations. One must be careful not to get too carried away with the theoretical aspects of economic analysis. Unless there is good reason to believe that substantial relative price changes are going to occur, it usually suffices to compare constant dollar annual cash flows of unequal life alternatives.

6.3   EXAMPLES

Example 6.1

Two machines that perform the same service have costs (at today's prices) as shown in the following list. Which machine is preferable if money costs 18% and inflation is 12%? There is no reason to expect any significant relative price changes for these machines and their operating costs:

	Machine A	Machine B
Initial cost	$20,000	$30,000
Annual operating cost	$7,000	$4,000
Life	5 yr	5 yr
Salvage value	0	0

Since both machines will serve for five years, it is proper to compare their annual cost for this time period. Let us do it first in constant dollar terms. There are no relative price changes, so the expenses stated represent constant dollar amounts in today's dollar. The real time value is

$$i_r = \frac{1.18}{1.12} - 1.0 = 5.36\%$$

The annual cash flow comparison is as follows:

$$ACF_A = \$7000 + \left(\frac{A}{P}, 5.36\%, 5\right) \cdot \$20,000 = \$11,665$$

$$ACF_B = \$4,000 + \left(\frac{A}{P}, 5.36\%, 5\right) \cdot \$30,000 = \$10,998$$

The alternatives are nearly equivalent, but machine B has a slightly lower cost.

The problem can also be solved in dollar terms but the solution is more tedious. First, the dollar cash flows must be determined:

	Constant Dollar Cash Flow			Dollar Cash Flow	
Time	Machine A	Machine B	$(1+f)^n$	Machine A	Machine B
0	20,000	30,000	1.00	20,000	30,000
1	7,000	4,000	1.12	7,840	4,480
2	7,000	4,000	1.25	8,781	5,018
3	7,000	4,000	1.40	9,834	5,620
4	7,000	4,000	1.57	11,015	6,294
5	7,000	4,000	1.76	12,336	7,049

Now the annual cash flow for each machine may be determined in dollar terms:

$$ACF_A = \left(\frac{A}{P}, 18\%, 5\right) \cdot [\$20{,}000 + \left(\frac{P}{F}, 18\%, 1\right) \cdot \$7{,}840$$

$$+ \left(\frac{P}{F}, 18\%, 2\right) \cdot \$8{,}781 + \left(\frac{P}{F}, 18\%, 3\right) \cdot \$9{,}834$$

$$+ \left(\frac{P}{F}, 18\%, 4\right) \cdot \$11{,}015 + \left(\frac{P}{F}, 18\%, 5\right) \cdot \$12{,}336]$$

$$= \$15{,}992$$

$$ACF_B = \left(\frac{A}{P}, 18\%, 5\right) \cdot [\$30{,}000 + \left(\frac{P}{F}, 18\%, 1\right) \cdot \$4{,}480$$

$$+ \left(\frac{P}{F}, 18\%, 2\right) \cdot \$5{,}018 + \left(\frac{P}{F}, 18\%, 3\right) \cdot \$5{,}620$$

$$+ \left(\frac{P}{F}, 18\%, 4\right) \cdot \$6{,}294 + \left(\frac{P}{F}, 18\%, 5\right) \cdot \$7{,}049]$$

$$= \$15{,}077$$

Again, the result is the same; machine B has a slightly lower cost. Note that whereas the annualized cash flows with dollar representation differ from those with constant dollar representation, the ratio between the annual costs is the same in both cases:

$$\frac{ACF_A}{ACF_B} = 1.061$$

This similarity shows that both methods give the same result.

Example 6.2

A private research firm has just started a new experimental project that is scheduled to last four years. A computer based data acquisition system is being purchased for the project. The equipment supplier has agreed to service the system at a guaranteed price of $90,000 per year for the four years, payable at the end of each year. The company is considering doing its own servicing. Equipment needed for diagnostics and servicing would cost about $150,000. A technician would also need to be hired at an estimated salary of $1583 per month plus $1417 per month of overhead and fringe benefits. The servicing equipment should have a salvage value to the company in four years of about one third the cost of new equipment. Review of the repairs required on similar equipment shows that about $3500 worth of parts are likely to be needed annually on the average.

Management would like to compare the annual cost of the company doing its own maintenance to the cost of $90,000 quoted by the supplier. The company has established an apparent time value of 18% and projects inflation to be 11%. No estimates have been made for future salary increases, computer parts prices, or the price of servicing equipment other than that they should increase at about the same rate as inflation.

The analysis time period is clearly four years here. It would be reasonable to compare the two alternatives in either dollar annual cash flows or constant dollar annual cash flows (or using present worth techniques). Since management wants to compare the cost of doing the servicing in-house to the $90,000 contract cost, it is probably best to calculate the dollar annual cash flow of the second option. However, it would be much easier to calculate the constant dollar annual cash flow since all the expenses are projected to stay constant in terms of constant dollars. The constant dollar annual cash flow can be calculated and then converted to a dollar annual cash flow by using Equation 6.5.

If today is used as the reference time, all expenses are already in constant dollars as stated. The yearly salary and parts cost is

$$12 \cdot [\$1583 + \$1417] + \$3500 = \$39,500$$

The real time value is

$$\frac{1.18}{1.11} - 1.0 = 0.063 = 6.3\%$$

The constant dollar annual cost is

$$ACF = \$39{,}500 + \left(\frac{A}{P}, 6.3\%, 4\right) \cdot \$150{,}000 - \left(\frac{A}{F}, 6.3\%, 4\right) \cdot SV$$

The salvage value was said to be one third the cost of new equipment. Presumably, this amount would be one third the cost of new equipment four years from now and not one third the dollar price now. If this assumption is true, the salvage value will be $50,000. Then

$$ACF = \$71{,}715$$

This figure cannot be compared to the contract cost until it is converted to a dollar annual cash flow. Using Equation 6.8, the calculation is

$$ACF_{\text{dollars}} = \$71{,}715 \frac{(A/P, 18\%, 4)\$}{(A/P, 6.3\%, 4)\overline{\$}} = \$91{,}733$$

The in-house option appears to be slightly more expensive than the contract.
   A note about the year end convention in this problem may be in order. The contract option had year end cash flows, and the in-house option had cash flows throughout each year. Some caution should be exercised when only one option requires application of the year end approximation. This caution is not due to inflation consideration but is good engineering economic analysis practice anytime. The year end convention in this case may tend to favor the in-house option since it would shift its cash flows to a later time on the average, but no such shifting is done for the contract option. Since the in-house alternative was the "loser," no further consideration is probably necessary.

Example 6.3

A worn-out conveyor belt in a food processing plant is being replaced. The belt is made of rubber on a fabric base. The belt typically must be replaced every two years. A more expensive belt built on a steel mesh base is available. It has an estimated life of five years. The fabric belt presently costs $1300 and the steel belt, $3500. The fabric belt has maintenance expenses of about $500 per year for dressing and minor repairs. The steel belt should have about half as much expense. Which belt is more economical? The company has established a cost of money of 16%, and inflation is about 12%.
   No information is given about the cost of replacement belts or maintenance

expense in the future. It is reasonable to assume that their prices will increase
at the same rate as inflation. Relative prices will stay constant then, and the
constant dollar cash flows for the first life cycle will be repeated every life
cycle. Therefore, it is reasonable to compare the two alternatives on a
constant dollar annual cash flow basis. The time value of money stated would
appear to be an apparent time value. The real time value is

$$i_r = \frac{1.16}{1.12} - 1.0 = 0.0357 = 3.57\%$$

If today is used as the reference time for the constant dollars, the annual cash
flows for the two alternatives are

$$ACF_{\text{fabric}} = \$500 + \left(\frac{A}{P}, 3.57\%, 2\right) \cdot \$1300$$

$$ACF_{\text{fabric}} = \$1185$$

$$ACF_{\text{steel}} = \$250 + \left(\frac{A}{P}, 3.57\%, 5\right) \cdot \$3500$$

$$ACF_{\text{steel}} = \$1027$$

The steel belt seems to be worth the extra initial expense.

It was noted in the text that simply stating cash flows in dollars and using
the apparent time value does not work for unequal life alternatives if the
annual cash flow is calculated for only one life cycle. It is interesting to see
what results would have been obtained if this incorrect calculation had been
used. Again assuming that relative prices stay constant, the dollar cash flows
for one life cycle of each alternative are as follows:

Year	Fabric Belt Constant Dollars	Dollars	Steel Belt Constant Dollars	Dollars
0	1300	1300	3000	3000
1	500	560	250	280
2	500	627	250	314
3			250	351
4			250	393
5			250	440

The dollar annual cash flow for one life cycle of the fabric belt is

$$ACF = \left(\frac{A}{P}, 16\%, 2\right) \cdot [\$1300 + \left(\frac{P}{F}, 16\%, 1\right) \$560$$

$$+ \left(\frac{P}{F}, 16\%, 2\right) \$627)]$$

$$= \$1400$$

The dollar annual cash flow for one life cycle of the steel belt is

$$ACF = \left(\frac{A}{P}, 16\%, 5\right)[\$3500 + \left(\frac{P}{F}, 16\%, 1\right) \cdot \$280$$

$$+ \left(\frac{P}{F}, 16\%, 2\right) \cdot \$314 + \left(\frac{P}{F}, 16\%, 3\right) \cdot \$351$$

$$+ \left(\frac{P}{F}, 16\%, 4\right) \cdot \$393 + \left(\frac{P}{F}, 16\%, 5\right) \cdot \$440]$$

$$= \$1413$$

This incorrect calculation shows the alternatives to be about equal. Although the results do not grossly differ on a relative basis from the correct calculations, there is a significant difference.

An incorrect calculation would also result if inflation had been totally ignored. The annual cash flows would then be calculated without adjusting the cash flows for inflation. The annual cash flow for the fabric belt in this incorrect calculation would be

$$ACF = \$500 + \left(\frac{A}{P}, 16\%, 2\right) \cdot \$1300 = \$1310$$

The corresponding annual cash flow for the steel belt is

$$ACF = \$250 + \left(\frac{A}{P}, 16\%, 5\right) \cdot \$3500 = \$1319$$

The relative error is about the same here as for the other incorrect calculation.

The errors due to using a dollar annual cash flow when actually constant dollar cash flows recur from life cycle to life cycle, and the errors due to ignoring inflation altogether are not extremely large in this example. Both lives are relatively short and inflation is fairly moderate. The errors would be greater if longer lives were involved or if inflation was greater.

CHAPTER SEVEN

Internal Rate of Return

7.1 RATE OF RETURN OF AN INVESTMENT

The internal rate of return calculation is probably the most commonly used analysis technique in engineering economics. It is also a technique that is easily misused and misinterpreted. Formally, the internal rate of return, hereafter referred to only as the rate of return, is the time value of money that results in a zero net present value for an investment. This formal definition is needed to be able to calculate the rate of return and to properly interpret the result. Conceptually, the rate of return is a measure of the profitability of an investment. The rate of return analysis method has two advantages over the present worth and annual cash flow methods. The calculation can be made independently of the time value of money. However, the time value must be specified before a formal decision can be made. The rate of return calculation gives not only a yes/no decision, but also a measure of the quality of an investment.

The rate of return calculation is illustrated in Figure 7.1. The typical investment as shown in Figure 7.2 consists of some initial expenditures that in turn result in increased income in the future. The larger the time value of money, the more the economic value of the future income is reduced relative to the expenditures that are required at an earlier point in time. Hence the negative slope of the curve in Figure 7.1 results. It is shown in Appendix 7.1 that investments with cash flows such as those in Figure 7.2 will result in the negative slope in Figure 7.1. Investments with complicated cash flows such as those in Figure 7.3 may produce present value curves that can result in multiple rate of return solutions. These cash flows do not nicely fit our concept of an investment of an expenditure now and return later. However, they can be solved by applying rate of return techniques, if properly handled. The discussion here will concentrate on the more typical investment type. However, the inflation related considerations presented later pertain equally well to the more complicated cash flow investments also.

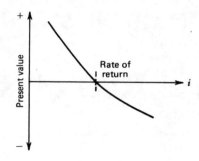

Figure 7.1 Determination of rate of return.

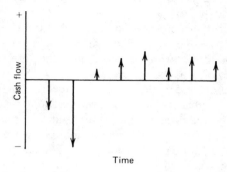

Figure 7.2 Typical cash flow for an investment.

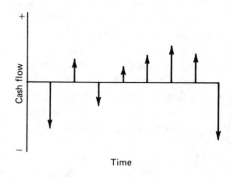

Figure 7.3 Nontypical cash flow for an investment.

Although the rate of return of an investment may be calculated independently of the time value of money, the interpretation of the results requires the time value to be established. The negative slope of the function

$$PV(i)$$

means that

$$R > i_t \Rightarrow PV(i_t) > 0 \qquad (7.1)$$

where R is the rate of return of the investment and i_t is the time value of money. Thus a rate of return greater than the time value of money implies that the present value is positive and also implies that the investment is economically desirable by the same criteria presented in Chapter 5. The rate of return calculation also gives a much better feel for the desirability of the investment than do the present values or annual cash flows. A rate of return of 19% means a lot more to the experienced engineering economist than does a present value of $59,300, for example. Engineering economics is not always the simple yes/no decision that investigation of the theory might lead one to believe. The rate of return of an investment often provides sufficient additional insight into the investment's economics to allow a sound decision to be made. Unfortunately, inflation can pose some real problems in the use of the rate of return in this less than rigorous manner.

Before going on to inflation considerations, the application of rate of return analysis in engineering economics problems needs a little more discussion. The rate of return indicates whether an investment of funds yields a satisfactory return. However, engineering economics most commonly deals with the comparison of mutually exclusive alternatives. One cannot simply calculate the rates of return of these alternatives and select the alternative with the largest value. In fact, many situations deal with alternatives where only costs exist and the rate of return of an individual alternative has no meaning. Most situations may be stated in terms of whether the extra investment required for a more expensive alternative results in enough savings or additional income to make it worthwhile. In this context the analysis of alternatives may be made to fit into our traditional concept of investment of funds and return on this investment.

The problem can be stated formally by letting

$$PV(i)_{net} = PV(i)_1 - PV(i)_2 \qquad (7.2)$$

where PV_1 is the present value of the more expensive alternative, PV_2 is the present value of the alternative requiring less investment, and PV_{net} is the net difference in present value of the two alternatives; PV_{net} is the same as the present value of the cash flows for alterantive 1 less the cash flows for alternative 2 on a year by year basis. These net cash flows will appear similar to those in Figure 7.2 for many comparisons, and $PV(i)_{net}$ will behave as shown in Figure 7.1. The rate of return calculated by setting PV_{net} to zero may be called the incremental rate of return for the more expensive alternative. By the reasoning presented earlier,

$$R_{net} > i_t \Rightarrow PV(i_t)_{net} > 0 \qquad (7.3)$$

That is, an incremental rate of return greater than the time value of money means that the difference in present value is positive. It follows directly from Equation 7.2 that

$$PV(i)_{net} > 0 \Rightarrow PV(i)_1 > PV(i)_2 \qquad (7.4)$$

Consequently,

$$R_{net} > i_t \Rightarrow PV(i_t)_1 > PV(i_t)_2 \qquad (7.5)$$

An incremental rate of return greater than the time value of money implies that the present value of the more expensive alternative is greater than the present value of the other alternative and is preferable by the criteria developed earlier. The converse conclusion is also true. An incremental rate of return less than the time value implies that the less expensive alternative has a greater present value.

With this rather lengthy review of the rate of return method now behind us, we can turn our attention to how inflation affects rate of return calculations.

7.2 DOLLAR AND CONSTANT DOLLAR CALCULATIONS

The use of dollars and the apparent time value or constant dollars and the real time value is no different with rate of return calculations from that with the economic analysis techniques discussed earlier. The derivations presented in Section 7.1 are equally valid with either method, but it is important to avoid confusion between the two methods. The rate of return will vary with the method used, but the time value is also different, and the decision remains the same. The rates of return calculated by using dollar cash flows are referred to as apparent rates of return. The rates of return calculated by using constant dollars are referred to as real rates of return. Real rates of return and real time values should be compared; likewise, apparent rates of return and apparent time values should be compared.

The real problem with inflation and rate of return analysis does not occur with formal analysis. As long as we follow the rules used with the other techniques, the decision remains the same. The problem arises in how the rates of return are interpreted. Many engineering economists are not familiar with the difference between dollars and constant dollars, let alone the difference between real and apparent rates of return. It is very easy to confuse real and apparent rates of return. Yet the two measures are as different as dollar and constant dollar annual cash flows. The real rate of return is less than the apparent rate of return by approximately the inflation rate. During periods of significant inflation an investment with a real rate of return of 5–10% may be attractive whereas an investment with an apparent rate of return of 10–15% will be undesirable. It is easy to see how confusion can result when people are used to referring only to rates of return and not stating whether they are real (based on constant dollar cash flows) or apparent (based on dollar cash flows).

When two alternatives are compared by applying the rate of return method, the incremental cash flow for the more expensive altenative is calculated. It is obviously necessary that the cash flows of both of these alternatives be stated in the same form, either dollars or constant dollars. The incremental cash flow for which the rate of return is calculated will then also be in this form. There is no easy way to have one alternative described in dollars and the other in constant dollars as was possible with the other analysis techniques. Once a rate of return is calculated, it may be converted from real to apparent or vice versa. Since the rate of return is simply the time value of money at which the present value is zero, Equation 3.15 will hold true for that time value as well as any other time value. Thus the relationship between real and apparent rate of return is

$$1 + R_r = \frac{1 + R_a}{1 + f} \qquad (7.6)$$

where R_r is the real rate of return and R_a is the apparent rate of return.* This relationship is seldom needed for the formal economic calculation as it is just as easy to change the time value to the other form if necessary. However, it is often a good idea to use Equation 7.6 to express the results in the form of the rate of return that has the most meaning. Managers who are used to accounting for inflation are likely to prefer the real rate of return figure as it

*A formal derivation is presented in Appendix 7.2.

gives a good indication of how the investment is performing. Managers who are not used to explicitly accounting for inflation are likely to prefer the apparent rate of return figure as it is how people are used to assessing investments and more closely realtes to how interest rates are reported in the money market.

7.3 EXAMPLES

Example 7.1

A family has saved about $2000 that they would now like to invest. There are a number of possibilities, of course, but the family has two particular possibilities that they would like to explore further. The first alternative is to insulate and weatherize their home. A cost analysis and a heat load analysis show that the family should be able to reduce their natural gas consumption from 250 MCF to 150 MCF (thousand cubic feet) for their $2000 investment. The second alternative is to buy a "mortgage contract" from a private individual who needs some cash. The contract has 10 years left and would provide year end payments of $325. The individual is willing to sell the contract for $2000.

The present price of natural gas is $2.80/MCF. The family expects to live in the house for about 10 more years. They do not expect the insulation to significantly affect resale value of the house. Which possibility provides the highest rate of return if inflation is about 9% over the next 10 years?

Let us consider the contract first. The cash flows with it are specified in dollars. Thus the apparent rate of return can be calculated directly:

$$0 = PV = -\$2000 + \left(\frac{P}{A}, i, 10\right) \$325$$

$$i = 10\% \qquad\qquad\qquad PV = -\$3.0$$
$$i = 9.5\% \qquad\qquad\qquad PV = +\$40.6$$
$$R_a = 9.97\%$$

The contract provides an apparent rate of return of 9.97% on the $2000 investment. The real rate of return can be calculated by converting the dollar cash flows to constant dollars. First, let us set t_0 to the present. Now the value of the $325 cash flows from the contract can be expressed in constant dollars:

Year	Dollar Cash Flow	Constant Dollar Cash Flow
0	−2000	−2000
1	+325	298
2	+325	274
3	+325	251
4	+325	230
5	+325	211
6	+325	194
7	+325	178
8	+325	163
9	+325	150
10	+325	137

The present value of the constant dollar cash flow can now be found:

$$i = 1\% \qquad\qquad PV = -\$9.2$$
$$i = 0.5\% \qquad\qquad PV = +\$36.9$$
$$R_r = 0.90\%$$

Although the contract provides an apparent rate of return of nearly 10%, it provides a real return of less than 1% with 9% inflation.

The insulation investment provides a 100 MCF savings in natural gas each year for the next 10 years. The price of this gas must be estimated for each year to determine the savings:

Year	Gas Price $[p'(t_j) \cdot p(t_0) \cdot (1 + f)^n]$	Dollar Cash Flow	Constant Dollar Cash Flow
0		−2000	−2000
1	$1.0 \cdot \$2.80/MCF \cdot$ $1.09 = \$3.05/MCF$	305	280
2	3.33	333	280
3	3.63	363	280
4	3.95	395	280
5	4.31	431	280
6	4.70	470	280

7	5.12	512	280
8	5.58	558	280
9	6.08	608	280
10	6.63	663	280

The dollar cash flow can be used to determine the apparent rate of return:

$$i = 16\% \qquad\qquad PV = \$20.8$$
$$i = 16.5\% \qquad\qquad PV = -\$21.0$$
$$R_a = 16.25\%$$

The constant dollar cash flow can be used to determine the real rate of return:

$$i = 7\% \qquad\qquad PV = -\$33.4$$
$$i = 6.5\% \qquad\qquad PV = \$12.9$$
$$R_r = 6.64\%$$

The insulation investment gives a real rate of return of 6.64% and an apparent rate of return of 16.25% with 9% inflation. It is clear that the insulation alternative provides a higher rate of return for the $2000 investment.

It would not have been necessary to evaluate both the real and apparent rates of return for both investments. One or the other would be sufficient. Also, once calculated in one form, the other form can be calculated by using Equation 7.6. It is important to note the big difference between the real and apparent rates of return even at this modest inflation rate. The real danger in this problem is that someone unaware of inflation would probably end up calculating an apparent rate of return for the mortgage since the cash flows are specified in dollars and at the same time calculate a real rate of return for the insulation since savings are easily calculated at today's prices. If these two rates of return were then compared, a totally incorrect indication of the relative value of the two investments would result. It is essential that real rates of return be compared only to other real rates of return and that apparent rates of return be compared only to other apparent rates of return.

Example 7.2

An old, heavily used warehouse currently has an incandescent lighting system. The lights run essentially 24 h/day, 365 days/yr and draw about 10 kW of power. Consideration is being given to replacing these lights with fluorescent lights to save on electricity. It is estimated that the same level of lighting can be achieved with 4.5 kW of fluorescent lights. Replacement of the lights will cost about $11,000. Bulb replacement and other maintenance are not expected to be significantly different. Electricity for the lights currently costs $0.049/kW·h. The relative electricity price is expected to increase about 3% annually for the next several years due to increasing fuel supply problems. The owners of the warehouse anticipate overall inflation to average about 9%. The warehouse is scheduled for demolition in five years to make way for a more modern facility. The company has an apparent time value of 14%. Should the company replace the incandescent lights with fluorescent lights?

Two alternatives exist here: keep the incandescent lights or install fluorescent lights. The selection could be made on the basis of the present cost or annual cash flow of each or by a rate of return comparison as is used here. The cash flows for the alternatives can be expressed in dollars or constant dollars. The constant dollar cash flows are used here.

The electricity used by the incandescent lights is

$$10 \text{ kW} \cdot 24 \ \frac{\text{hr}}{\text{day}} \cdot 365 \ \frac{\text{day}}{\text{yr}} = 87,600 \text{ kW} \cdot \text{h/yr}$$

The electricity that would be used by the fluorescent lights is

$$4.5 \text{ kW} \cdot 24 \ \frac{\text{hr}}{\text{day}} \cdot 365 \ \frac{\text{day}}{\text{yr}} = 39,420 \text{ kW} \cdot \text{h/yr}$$

The constant dollar cash flows for each alternative are then as follows, using the present as the reference time:

Year	p_{elec}^r	Incandescent ($)	Fluorescent ($)
0	1.00	—	11,000
1	1.030	4421	1,990
2	1.061	4554	2,049
3	1.093	4690	2,111
4	1.126	4831	2,174
5	1.159	4976	2,239

The fluorescent lights require the extra initial investment in this case. The incremental cash flow for this investment can be found by subtracting the cash flows of the fluorescent lights from the cash flows for the incandescent lights:

Year	Incremental Cash Flow for Fluorescent Lights ($)
0	−11,000
1	2,431
2	2,504
3	2,579
4	2,656
5	2,737

The rate of return for this cash flow can now be found. Let

$$i = 5\% \qquad PV = +\$144$$
$$i = 6\% \qquad PV = -\$164$$
$$i = 5.5\% \qquad PV = -\$11 \approx 0$$

The real rate of return on the expense required to put in the fluorescent lights is 5.5%. This return can be converted to an apparent rate of return:

$$R_a = 1.055 \cdot 1.09 - 1.0 = 15.0\%$$

The apparent rate of return is a little higher than the apparent time value of money, and thus the fluorescent lights appear to be worthwhile.

This problem could just as well have been solved in dollar terms to obtain the apparent rate of return directly. Also, the apparent time value could have been converted to a real time value for comparison to the real rate of return.

APPENDIX 7.1   EVALUATION OF $\partial PV/\partial i$

The objective of this appendix is to show that the derivative

$$\frac{\partial PV}{\partial i}$$

is negative for sets of cash flows that are of the type shown in Figure 7.2. The present value of a single cash flow is

$$P = F \cdot [1 + i]^{-j} \qquad\qquad \text{(A7.1-1)}$$

where P is the present value of the single cash flow F and

$$j = \frac{t - t_0}{\Delta t}$$

The effect of changing i on the present value for this single cash flow is

$$\frac{\partial P}{\partial i} = F \cdot [1 + i]^{-j-1} \cdot [-j] \qquad\qquad \text{(A7.1-2)}$$

The relative change in present value due to a change in i can be found by dividing Equation A7.1-2 through by the present value:

$$\frac{[\partial P/\partial i]}{P} = \frac{-j}{1 + i} \qquad\qquad \text{(A7.1-3)}$$

The relative change will always be negative as long as $i > -1$ and $j > 0$, conditions that are easily met. Also, it is seen that the relative effect of a change in i on the present value is proportional to $-j$. There is more and more relative effect on the magnitude of the present value as j gets larger.

Consider a set of cash flows such as shown in Figure 7.2. The cash flows can be divided into two groups. Group 1 consists of the negative cash flows and group 2, the positive cash flows. Let us use $-$ to denote the first group and $+$ to denote the second group. All cash flows in the first group occur before any cash flows in the second group. From Equation A7.1-3, it is evident that

$$\frac{\partial PV^+/\partial i}{PV^+} < \frac{\partial PV^-/\partial i}{PV^-} \qquad\qquad \text{(A7.1-4)}$$

Since both sides of this inequality are negative, it follows that

$$\frac{|\partial PV^+/\partial i|}{|PV^+|} > \frac{|\partial PV^-/\partial i|}{|PV^{-1}|} \qquad\qquad \text{(A7.1-5)}$$

or

$$\frac{|\,\partial PV^+/\partial i\,|}{|\,\partial PV^-/\partial i\,|} > \frac{|\,PV^+\,|}{|\,PV^-\,|} \qquad (\text{A7.1--6})$$

The net present value for the cash flows is

$$PV_{\text{net}} = PV^+ + PV^-$$

and the derivative with respect to i is

$$\frac{\partial PV_{\text{net}}}{\partial i} = \frac{\partial PV^+}{\partial i} + \frac{\partial PV^-}{\partial i} \qquad (\text{A7.1--7})$$

As long as

$$PV_{\text{net}} \geq 0$$

it is seen from Inequality A7.1–6 that

$$\left|\frac{\partial PV^+}{\partial i}\right| > \left|\frac{\partial PV^-}{\partial i}\right| \qquad (\text{A7.1--8})$$

and from Equation A7.1–2, it is evident the derivative for the positive group is negative and is positive for the negative group. Combining this information with Equation A7.1–7, it follows that

$$\frac{\partial PV_{\text{net}}}{\partial i} < 0$$

when

$$PV_{\text{net}} \geq 0$$

Thus we may conclude for this type of set of cash flows that the slope of $PV(i)$ is negative when it is above the zero axis as shown in Figure 7.1 and will cross the axis with a negative slope. It is possible for the slope to become positive once the function is below the zero axis, but it can never cross the axis or even touch it again since the slope would be negative when it reached the zero axis.

APPENDIX 7.2   RELATIONSHIP BETWEEN REAL AND APPARENT RATE OF RETURN

The relationship for the real rate of return is

$$0 = \sum_{j=0}^{n} Y_j^c \cdot [1 + R_r]^{-j} \tag{A7.2-1}$$

where Y^c represents the constant dollar cash flows of the investment and the equation is solved for R_r, the real rate of return. The corresponding relationship for the apparent rate of return is

$$0 = \sum_{j=0}^{n} Y_j^d \cdot [1 + R_a]^{-j} \tag{A7.2-2}$$

where Y^d represents the dollar cash flows for the same investment and R_a is the apparent rate of return. The dollar cash flows and the constant dollar cash flows are related to each other by

$$Y_j^d = Y_j^c \frac{\bar{p}(t_j)}{\bar{p}(t_0)} \tag{A7.2-3}$$

For a constant rate of inflation f, we obtain

$$Y_j^d = Y_j^c \cdot \frac{\bar{p}(t_{j=0})}{\bar{p}(t_0)} \cdot [1 + f]^j \tag{A7.2-4}$$

This relationship may be substituted into Equation A7.2-2:

$$0 = \sum_{j=0}^{n} Y_j^c \cdot \frac{[1 + R_a]^{-j}}{[1 + f]^{-j}} \tag{A7.2-5}$$

Comparing Equations A7.2-5 and A7.2-1, it is seen that

$$[1 + R_a] = [1 + R_r] \cdot [1 + f] \tag{A7.2-6}$$

must hold true since the Y_j^c are the same in both equations and R_a, R_r, and f are constants.

CHAPTER EIGHT

Taxes

No new engineering economic analysis theory is needed to deal with taxes. Taxes are just another expense. They should be treated no differently from any other expense. Inflation has an impact on taxes that must be closely observed in many cases; thus a closer look at this subject is warranted. Only U.S. federal income taxes are discussed here; however, the principles apply to other taxes as well.

It must be realized that tax laws are continually being changed. The discussions that follow are intended to reflect the tax code as it exists for the 1980 tax year. Undoubtedly, a number of changes will be made in the next few years. Lawmakers appear to be beginning to deal with some of the tax code problems that inflation casuses. For example, some interest payments will be exempt from taxes in the 1981 tax year, and the new administration has proposed to index tax brackets to inflation. The discussion that follows shows the kinds of considerations that must be made when calculating taxes for engineering economic analyses. However, it is important that the engineering economist be aware of the changes that will undoubtedly take place in the years to come.

8.1 TAX CALCULATION

The tax laws in the United States are incredibly complex. Obviously, a complete review of tax laws cannot be given here. Only a few general principles of the tax structures are discussed to demonstrate how inflation can be accounted for in determining tax expenses. Income tax is based on the taxable income, which, for corporations, is

$$TI = G - E - D \qquad (8.1)$$

where TI is the taxable income, G is the gross income, E is the expense, and

TABLE 8.1 Tax Rates for Corporations (1980 Tax Year)

Taxable Income ($)	Tax Is: This Amount ($)	Plus This (1%)	Of *TI* over ($)
$TI \leq 25{,}000$	0	17	0
$25{,}000 < TI \leq 50{,}000$	4,250	20	25,000
$50{,}000 < TI \leq 75{,}000$	9,250	30	50,000
$75{,}000 < TI \leq 100{,}000$	16,750	40	75,000
$TI > 100{,}000$	26,750	46	100,000

D is the depreciation. Gross income pertains essentially to all income except for the sale of capital assets (equipment and real estate used to produce income). Expense pertains to the expenses incurred in generating the income, again except for the expense of capital assets. Depreciation is an allowance for the cost of capital assets. For individuals, the adjusted gross income is

TABLE 8.2 Tax Rates for Single Taxpayers (1980 Tax Year)

Taxable Income ($)	Tax Is: This Amount ($)	Plus This (%)	Of *TI* over ($)
$TI \leq \$2{,}300$	0	0	0
$2{,}300 < TI \leq 3{,}400$	0	14	2,300
$3{,}400 < TI \leq 4{,}400$	154	16	3,400
$4{,}400 < TI \leq 6{,}500$	314	18	4,400
$6{,}500 < TI \leq 8{,}500$	692	19	6,500
$8{,}500 < TI \leq 10{,}800$	1,072	21	8,500
$10{,}800 < TI \leq 12{,}900$	1,555	24	10,800
$12{,}900 < TI \leq 15{,}000$	2,059	26	12,900
$15{,}000 < TI \leq 18{,}200$	2,605	30	15,000
$18{,}200 < TI \leq 23{,}500$	3,565	34	18,200
$23{,}500 < TI \leq 28{,}800$	5,367	39	23,500
$28{,}800 < TI \leq 34{,}100$	7,434	44	28,800
$34{,}100 < TI \leq 41{,}500$	9,766	49	34,100
$41{,}500 < TI \leq 55{,}300$	13,392	55	41,500
$55{,}300 < TI \leq 81{,}800$	20,982	63	55,300
$81{,}800 < TI \leq 108{,}300$	37,677	68	81,800
$TI > 108{,}300$	55,697	70	108,300

TABLE 8.3   Tax Rates for Married Taxpayers Filing Jointly (1980 Tax Year)

Taxable Income ($)	Tax Is This Amount ($)	Plus This (%)	Of *TI* over ($)
$TI \leq \$3,400$	0	0	0
$3,400 < TI \leq 5,500$	0	14	3,400
$5,500 < TI \leq 7,600$	294	16	5,500
$7,600 < TI \leq 11,900$	630	18	7,600
$11,900 < TI \leq 16,000$	1,404	21	11,900
$16,000 < TI \leq 20,200$	2,265	24	16,000
$20,200 < TI \leq 24,600$	3,273	28	20,200
$24,600 < TI \leq 29,900$	4,505	32	24,600
$29,900 < TI \leq 35,200$	6,201	37	29,900
$35,200 < TI \leq 45,800$	8,162	43	35,200
$45,800 < TI \leq 60,000$	12,720	49	45,800
$60,000 < TI \leq 85,600$	19,678	54	60,000
$85,600 < TI \leq 109,400$	33,502	59	85,600
$109,400 < TI \leq 162,400$	47,544	64	109,400
$162,400 < TI \leq 215,400$	81,464	68	162,400
$TI > 215,400$	117,504	70	215,400

used instead of the gross income. The adjusted gross income reflects a $1000 exemption for each dependent plus other deductions for some people.

Once the taxable income is determined, the income tax is calculated according to the appropriate set of tax rates. Tax rates for corporations are shown in Table 8.1. Tax rates for single taxpayers are shown in Table 8.2 and for married taxpayers filing jointly, in Table 8.3. Most taxpayers fall into one of these three categories.

The most important feature of income taxes to remember is that they are based on dollar profits and dollar income and are paid in dollars. No consideration is given to the effect that inflation has on the value of the dollar. Consequently, it is often necessary to first calculate taxes in dollar terms even though a particular engineering economics problem is to be solved in constant dollar terms. This requirement is particularly important when the economics pertain to an individual rather than a corporation. The reason for the dollar calculation is that the effective tax rate varies with dollar income rather than with constant dollar income. If overall taxed income is sufficiently large that the tax rates no longer increase with increasing income ($100,000 for corporations, $108,302 for single taxpayers, and $215,400 for married taxpayers), all additional income is taxed at the same rate (46%

for corporations, 70% for individuals), and taxable income may be calculated in constant dollars and taxes calculated in constant dollars. Most large corporations easily fall into this category; few individuals do. This direct calculation of constant dollar taxes is only possible because dollar taxes are

$$T_I^d = TI^d \cdot r \qquad (8.2)$$

where T_I^d is the income tax in dollars, TI^d is the taxable income in dollars, and r is the effective tax rate. The tax in constant dollar terms is

$$T_I^c = T_I^d \; \frac{\bar{p}(t_0)}{\bar{p}(t)} \qquad (8.3)$$

where t is the year of the tax calculation. The taxable income in constant dollar terms is

$$TI^c = TI^d \; \frac{\bar{p}(t_0)}{\bar{p}(t)} \qquad (8.4)$$

and thus the resulting constant dollar income tax is

$$T_I^c = TI^c \cdot r \qquad (8.5)$$

There is still a slight approximation involved in Equation 8.5 as there is no allowance for the changing value of a dollar during the tax year. The actual tax calculation is analogous to applying year end convention to dollar cash flows. Equations 8.3 and 8.4 properly convert these year end dollar cash flows to constant dollars. Equation 8.5 implies that constant dollar cash flows throughout the year may be summed to get the taxable income for the year. The discrepancy is similar to the discrepancy that can result from inconsistent application of the year end convention discussed in Chapter 4. The resulting error is usually minor compared to other approximations that must be made in economic analysis. Where a high level of precision is necessary, it is usually best to work in dollar terms, at least for the tax calculations. Some tax calculations such as income averaging or carrying forward and carrying back of taxes may be based on dollar cash flows that span several years. These tax calculations should all be made in dollar terms to avoid errors. Also, depreciation and capital gains involve dollar cash flows that may span many years. This particular problem is discussed in more detail a little later.

For corporations and individuals who do not have sufficient income to have a constant tax rate, there is no substitute for calculating taxable income in dollars and determining the required tax from the appropriate tax table. The effective tax rate increases with increasing dollar income. The constant dollar amount of income taxes will increase with time as a result of inflation even though the constant dollar cash flows remain constant. The increasing dollar cash flows that result from the inflation and not increased value of the transactions steadily push the individual or corporation into tax brackets that have increasingly higher tax rates. This increase in tax rate decreases the real return on most investments. It also may prevent cash flows from recurring from one life cycle to the next for some equipment, if the constant dollar taxes increase from one life cycle to the next. This result can complicate the comparison of unequal life alternatives as was discussed in Chapter 6.

8.2 DEPRECIATION AND CAPITAL GAINS

Depreciation is a means to account for the cost of a capital asset in determining taxable income. A part of the initial cost of the asset is subtracted from the income each year. This allowance, or depreciation, is calculated in a number of different ways depending on the regulations for a particular item. Two of the more common techniques are straight-line and sum-of-the-year's digits.

The straight-line method of calculating depreciation allows the depreciation to be the same amount each year. Mathematically, it is

$$d_j = \frac{IC - SV}{n} \qquad j = 1, n \tag{8.6}$$

where d_j is the depreciation for the jth year, IC is the initial cost, SV is the salvage value allowed for tax purposes, and n is the life set for tax purposes. The amount d is subtracted from the income each year to account for the cost of the capital asset. However, the total depreciation that can be claimed is the initial cost less any salvage value. No more depreciation may be claimed after the nth year.

The sum-of-the-year's digits method of depreciation allows more depreciation at first but with less in later years. The depreciation is calculated as follows:

$$d_j = \frac{n + 1 - j}{SYD} \cdot [IC - SV] \tag{8.7}$$

where

$$SYD = \sum_{j=1}^{n} j$$

Other depreciation techniques are also used, and the engineering economist should be familiar with them. However, they are not described here, since the particular depreciation technique used is not important for the following discussion.

The important point to note is that depreciation is based on initial dollar cost. The amount of depreciation allowed in the years that follow the purchase of the asset continue to be based on the original dollar cost even though the value of the dollar changes. Since the depreciation is claimed for a number of years after the initial purchase, the value of depreciation claimed may be much less than the actual value of the asset. The depreciation is calculated in dollars less valuable than the original purchase. This problem becomes painfully clear when the company replaces the asset and finds that it costs much more than the original cost, some years earlier, of the one replaced.

This requirement of basing depreciation on dollar costs means that there is no choice but to express it in dollar terms in an economic analysis. If an analysis is being conducted in constant dollar terms, it is usually necessary to calculate the depreciation first in dollar terms and then convert it to constant dollars.

A variable called the book value of an asset must be determined for tax calculations. The book value is simply the initial cost less any depreciation that has been claimed. Again, the book value is calculated only in dollar terms and does not account for inflation. When an asset is sold, the difference between the selling price and the book value is called the capital gain. That is, if the asset is sold for more than the book value, the difference is a capital gain, and if it is sold for less than the book value, the difference is a capital loss. Capital gains are taxable. The calculation of capital gains taxes can be rather involved. However, for corporations with more than $50,000 of taxable income, they are taxed at a rate of 28%.

Again, the capital gain is calculated strictly on the basis of the dollar initial cost and the dollar depreciation. The selling price of the asset in value terms may be less than the initial cost less depreciation in value terms, meaning that there is a real capital loss, yet the tax calculation can still show a net capital gain since it does not account for the changing value of the dollar.

The end result of the way depreciation and capital gains must be calculated is to increase the value of the taxes that must be paid when the inflation rate is high. Inflation causes the depreciation to be less than the amount that truly represents the value of the capital asset, and it causes the capital gain to be larger than it should be in value terms. The larger taxes with high inflation rates means that the net return on capital investment is decreased by inflation even if the return before taxes in value terms is not affected by inflation.

8.3   PAPER INVESTMENTS AND BORROWING

Inflation has an especially detrimental effect on paper investments such as corporate bonds, savings accounts, and certificates of deposit. Interest payments for these investments are based on the dollar principle, and of course, are paid in dollars. Interest is considered taxable income, and tax is due on the interest payments. While tax is being paid on all of the interest collected, the principle is declining in value due to inflation. That is, tax is paid on the apparent return of a paper investment not its real return. It is common for paper investments to have a net negative yield since the interest payments are often insufficient to offset the declining real value of the principle due to inflation. Tax is still paid on the interest earned, however.

Previous discussion and examples have shown how important it is to differentiate between value or constant dollar returns and dollar returns. This distinction is especially important for paper investments since not only is the return set in dollar terms, but taxes must be paid on a return that may not exist in real terms. With most investments, dollar expenses and dollar returns tend to increase together as a result of inflation. The taxable income reflects both the increased expense and income and tends to stay more or less constant in value terms. There is an "inflation expense" associated with a paper investment from the lost purchasing power of the principle. This lost purchasing power is due to inflation. Interest rates may rise to reflect this "expense," but the expense is not allowed in the tax codes. Thus paper investments are increasingly affected by inflation even when interest rates increase to reflect inflation.

Calculation of the return for paper investments is no different from that for other investments. It just must be remembered that interest is based on the dollar principle, interest is paid in dollars, taxes are based on the dollar interest payments, and the principle is repaid in dollars. A simple relationship between the stated interest rate and the real after tax rate of return may be derived for paper investments that have year end interest and tax payments. The interest payment is

$$I_{bt} = i_s \cdot S \qquad\qquad (8.8)$$

where I_{bt} is the interest payment, i_s is the stated interest rate, and S is the principle. The interest remaining after tax is

$$I_{at} = [1 - r_e] \cdot I_{bt} \tag{8.9}$$

where I_{at} is the interest left after the income tax is paid and r_e is the effective tax rate. The dollar cash flows for the investment are then $-S$ at time zero and $S + I_{at}$ one year later. The apparent rate of return is found by setting the present value of these cash flows to zero:

$$0 = -S + \frac{S + I_{at}}{1 + R_a} \tag{8.10}$$

or

$$R_a = \frac{I_{at}}{S} \tag{8.11}$$

Substituting from Equations 8.8 and 8.9 for I_{at} gives

$$R_a = [1 - r_e] \cdot i_s \tag{8.12}$$

Converting this result to a real rate of return gives

$$R_r = \frac{1 + [1 - r_e] \cdot i_s}{1 + f} - 1 \tag{8.13}$$

This equation is valid only if i_s, r_e and f are all constant and if interest and taxes are paid only at the end of the year. However, it does give reasonable approximations for most other situations also. It is seen that the real return on paper investments will be substantially less than the stated interest if there is significant inflation or a sizable tax rate.

What is true for paper investments is true in reverse for borrowers. The interest paid on borrowed money is based on the dollar principle, interest payments are considered as expenses for tax calculations, and the principle is repaid in dollars that are of less value than the original loan. All the relationships derived here for paper investments also apply to borrowing. Equation 8.13 describes the real cost in value terms of borrowing money. Inflation and taxes tend to make borrowing less costly than the stated interest would imply. Some caution should be expressed in determining the advisability of borrowing, however. The low real cost that may exist is not a guarantee of profitability unless it can be used for investments that have a higher real return after tax.

8.4 EXAMPLES

Example 8.1

A recent engineering graduate accepted a job offer for $25,800 annually. The engineer is married and has one child and will utilize the standard deduction (already built into the tax table). How much federal income tax will the engineer be required to pay? If inflation averages 12% and the engineer also receives 12% pay raises every year, what will the dollar income and dollar income tax be in 10 years? What is the average income tax rate the engineer pays now and in 10 years? What is the buying power of the after-tax income 10 years from now in today's dollar? What average annual pay increase would this engineer have to receive for the next 10 years to have the same after-tax income in today's dollar 10 years from now? Assume that the tax structure does not change during the next 10 years.

There are many questions here, so let us take them one at a time. The adjusted gross income for the engineer should be

$$\$25,800 - 3 \cdot \$1000 = \$22,800$$

since the engineer has three exemptions and we assume the family has no other income. The tax, from Table 8.3, is

$$\$3273 + 0.28 \cdot [\$22,800 - \$20,200] = \$4001$$

The engineer would pay $4001 in federal income tax.

Ten years of 12% annual pay raises would result in an income of

$$\$25,800[1.12]^{10} = \$80,131$$

The taxable income would be $80,131 - 3 \cdot \$1000 = \$77,131$. The income tax on this taxable income would be

$$\$19,678 + 0.54 \cdot [\$77,131 - \$60,000] = \$28,929$$

The average income tax rate would be the total income tax divided by the total income. The average rate now is

$$\frac{4,001}{25,800} = 0.155$$

or 15.5% of the engineer's income would be used to pay federal income tax.

Ten years from now the average rate would be

$$\frac{28,929}{80,131} = 0.361$$

and 36.1% of the engineer's income would be used for federal income tax.

The buying power of the income is the same as its constant dollar value. The after-tax dollar income in 10 years is

$$\$80,131 - \$28,929 = \$51,202$$

The value of this after-tax income in today's dollar is

$$\$51,202 \cdot [1.12]^{-10}\$/\$ = \$16,486$$

The engineer presently has an after-tax income of

$$\$25,800 - \$4,001 = \$21,799$$

Even though pay raises commensurate with inflation are given the engineer, the buying power of the after-tax income decreases about 25% in 10 years.

Determination of the annual pay raise required for the engineer to stay even is a little more difficult. The calculations are the same as those already done here, but a trial and error solution is required. Let us start by assuming a 15% annual pay raise is required. Income in 10 years would be

$$\$25,800 \cdot [1.15]^{10} = \$104,375$$

Income tax would be

$$\$33,502 + 0.59 \cdot [\$101,375 - \$85,600] = \$42,809$$

After tax income would be

$$\$104,375 - \$42,809 = \$61,566$$

In today's dollar, this amount is

$$\$61,566 \cdot [1.12]^{-10}\$/\$ = \$19,822$$

This is still less than the present after-tax income of $21,799. Similar calculations can be made for different guesses:

Annual Pay Raise (%)	After-Tax Income in Today's Dollar ($)
12	16,486
15	19,822
16	21,046
17	22,228

A linear interpolation between the last two numbers should provide a good estimate:

$$16\% + \frac{\$21,799 - \$21,046}{\$22,228 - \$21,046} \cdot 1\% = 16.6\%$$

This engineer would need a 16.6% annual pay raise for the next 10 years just to maintain the same level of after-tax buying power even though inflation is only 12%.

Note that state and local taxes are not considered in any of these calculations. They will alter the outcome somewhat but will not change the overall conclusions. Also, note that it is assumed that the tax rates do not change. It is probably quite likely that if inflation averages 12% for 10 years in the United States, there would be some modification of the tax rates. However, if history is a guide, it is likely that the tax structure would not be changed fast enough to fully correct for inflation. An interesting exercise is to determine how to modify the tax structure to keep income taxes in constant dollar terms the same. It is not done simply by decreasing tax rates at the same rate as inflation increases.

Example 8.2

A company is considering the installation of a heat regenerator on an office building. The regenerator uses the energy in the exhaust air to temper the fresh air, saving energy for heating and cooling. The regenerator will cost about $100,000 and will save about $20,000 annually in fuel cost, both at today's prices. The price of energy is expected to increase at the same rate as inflation. The company has an effective income tax rate of 48% including state and local income tax. The regenerator has an expected life of 10 years and no salvage value. Determine the real rate of return for this investment when there is no inflation and when there is 12% inflation. Use straight-line depreciation. Ignore investment tax credits and any other special tax credits for this item.

Probably the simplest way to solve this problem is to calculate everything in dollar terms, solve for the apparent rate of return and convert it to the real rate of return. The dollar depreciation will be the same in either case. Assuming that the tax life is the same as the economic life, the depreciation is

$$d = \frac{\$100,000 - 0}{10} = \$10,000$$

The no inflation case is the easiest to solve since the cash flows are the same each year. The taxable income generated by the investment each year is

$$TI = \$20,000 - 0 - \$10,000 = \$10,000$$

The income tax due is then

$$TI = \$10,000 \cdot 48\% = \$4800$$

The net cash flow each year is

$$Y = \$20,000 - \$4800 = \$15,200$$

Since there is no inflation, this cash flow may be considered as either dollars or constant dollars, there is no difference. The rate of return may now be calculated:

$$0 = PV = -\$100,000 + \left(\frac{P}{A}, i, 10\right) \cdot \$15,200$$

Let

$$i = 10\%, \quad PV = -\$6,600$$
$$i = 5\%, \quad PV = +\$17,370$$
$$i = 8.5\%, \quad PV = -\$270 \approx 0$$

The investment earns a return of about 8.5%. Since there is no inflation here, this return is both the real and the apparent return.

The 12% inflation situation is a little more difficult. The depreciation stays the same in dollar terms, but the savings will increase as the cost of energy increases. The savings each year are:

Year	$\dfrac{\bar{p}(t)}{\bar{p}(0)}$	Savings ($)
1	1.120	22,400
2	1.254	25,100
3	1.405	29,000
4	1.574	31,500
5	1.762	35,200
6	1.974	39,500
7	2.211	44,200
8	2.476	49,500
9	2.773	55,500
10	3.106	62,100

The savings are calculated here using the year end convention described in Chapter 4. These same figures are also used for the tax calculations in the list that follows. It should be realized, however, that some error will result in the tax calculations since the actual dollar cash flows on which the savings are based are slightly smaller. The resulting error will not be large.

The taxable income income and net cash flow may be calculated for each year:

Year	Depreciation ($)	*TI* ($)	Tax ($)	Net Cash Flow ($)
0	—	—	—	−100,000
1	10,000	12,400	5,950	16,450
2	10,000	15,100	7,250	17,850
3	10,000	19,000	9,120	19,880
4	10,000	21,500	10,320	21,180
5	10,000	25,200	12,100	23,100
6	10,000	29,500	14,160	25,340
7	10,000	34,200	16,420	27,780
8	10,000	39,500	18,960	30,540
9	10,000	45,500	21,840	33,660
10	10,000	52,100	25,010	37,090

The apparent rate of return for the net dollar cash flows can now be calculated, let

$$i = 15\%, \quad PV = \$14,600$$
$$i = 20\%, \quad PV = -\$7,040$$
$$i = 18.2\%, \quad PV = -40 \approx 0$$

The investment earns an apparent rate of return of 18.2% with the 12% inflation. However, it is the real rate of return that is important for comparison purposes. The real rate of return is

$$R_r = \frac{1.182}{1.12} - 1.0 = 0.055 = 5.5\%$$

The investment earns a real rate of return of 5.5% with 12% inflation as compared to 8.5% with no inflation. This difference is due only to the fact that the depreciation allowance does not account for inflation. The savings in value terms was the same before taxes in both cases.

Example 8.3

With an inflation rate of 12%, what stated interest rate is required for a real return or cost of zero for paper investments or borrowing for a corporation which has substantial income?

The federal tax rate for such a corporation is 46%. The required interest for a net rate of zero can be found by setting Equation 8.13 to zero:

$$0 = \frac{1 + [1 - r_e] \cdot i_s}{1 + f} - 1$$

$$i_s = \frac{f}{1 - r_e}$$

With 12% inflation and a 46% tax rate, the result is

$$i_s = \frac{0.12}{0.54} = 0.22 = 22\%$$

Inflation and taxes combine to result in a 22% stated interest rate, giving a net return or cost of zero. State and local income taxes may make this amount even larger.

Example 8.4

An individual in the 37% tax bracket purchased an item for $1000 on an installment plan with 12 monthly payments and a stated interest rate of 1½% each month. If inflation averages 11% over this one year period, what is the real cost of this loan? Assume the 12 month time period approximately coincides with the tax year.

The 12 monthly payments are

$$A = \$1,000 \cdot \left(\frac{A}{P}, 1\tfrac{1}{2}\%, 12 \right) = \$91.68$$

The portions of each payment that are interest and principle are as follows:

Month of Payment	Interest ($)	Principle ($)	Loan Balance ($)
1	15.00	76.68	923.32
2	13.85	77.83	845.49
3	12.68	79.00	766.49
4	11.50	80.18	686.31
5	10.29	81.39	604.92
6	9.07	82.61	522.32
7	7.83	83.85	438.47
8	6.58	85.10	353.37
9	5.30	86.38	266.99
10	4.00	87.68	179.32
11	2.69	88.99	90.32
12	1.35	90.32	0.00
Total	100.14	1000.00	

The after-tax dollar cash flows for the loan are then

Month:	0	1	2	· · ·	11	12
Cash flow:	+$1000	−$91.68	−$91.68	· · ·	−$91.68	−$91.68 + 0.37 · $100.14

The after-tax dollar cost of this loan is found in the same manner as the rate of return:

$$0 = PV = \$1000 - \left(\frac{P}{A}, i, 12\right) \cdot \$91.68 + \left(\frac{P}{F}, i, 12\right) \cdot \$37.05$$

where i is the return with a monthly compounding period. The equation may now be solved. Let

$$i = 1.0\%, \quad PV = +\$1.01$$
$$i = 0.9\%, \quad PV = -\$5.14$$
$$i = 0.984\%, \quad PV = \$0.03 \approx 0$$

The apparent cost of the loan after tax is 0.984% per month. The real cost after tax can be calculated, but first the equivalent monthly inflation rate must be determined by using Equation A2.1-4:

$$f = (1.11)^{1/12} - 1.0 = 0.00873 = 0.873\%$$

where f is the monthly inflation rate. The real after-tax interest rate is then

$$R_r = \frac{1.00984}{1.00873} - 1.0 = 0.0011 = 0.11\%$$

The loan has a real cost of 0.11% per month rather than the stated cost of 1½%. The stated 1½% per month rate is equivalent to the following annual interest:

$$(1.015)^{12/1} - 1.0 = 0.196 = 19.6\%$$

The real after-tax cost can also be expressed in annual terms:

$$(1.0011)^{12/1} - 1.0 = 0.0132 = 1.32\%$$

CHAPTER NINE

Sensitivity and Uncertainty

Engineering economics deals with the future, of which we can never be certain. Engineering economic theory is oriented toward yes/no decisions based on a given set of conditions. Real economic decisions are seldom that simple. One or more of the variables in an economic analysis may contain considerable uncertainty, and a single evaluation of economics with some expected values of these variables seldom provides sufficient information on which to base a wise decision. It is much better to know what the outcome will be over the possible range of such variables. Then a decision can be made that includes consideration of the consequences when important variables are different from those expected.

Sensitivity analysis is defined here as the determination of the economic performance of engineering alternatives as measured by present value, annual cash flow or rate of return, and as a function of one or more variables. A sensitivity analysis is used to evaluate the effects of uncertainty. Sensitivity analysis is also used in more formal risk analysis where probabilities may be assigned to possible values of variables. The following discussion concentrates on sensitivity analysis and does not deal with the distinctions between uncertainty and risk.

9.1 INFLATION AS A FACTOR IN SENSITIVITY ANALYSIS

Sensitivity analyses are commonly performed on variables such as throughput, utilization, life, and price of a product. These types of variables are often the dominant factor in determining the economic outcome of an investment. Any variable that is not fixed and that may alter the economic outcome is a potential candidate for sensitivity analysis. Inflation is not normally included as a factor in sensitivity analysis, but it often should be. It was seen in Chapter 8 that simply as a result of taxes the return on an investment can be very significantly affected by inflation. Even investments that were not affected by inflation before taxes were significantly affected by inflation after taxes. Other investments may be directly affected by inflation, particularly where contracts or other fixed terms are involved. Furthermore, the value of future inflation is very uncertain. Predictions of the inflation rate for next year vary by as much as 10 percentage points. Predictions of inflation for the next five to ten years are very speculative, indeed. Thus any economic analysis that is even moderately sensitive to inflation should include a sensitivity analysis for inflation.

Even when inflation is not an explicit variable in a sensitivity analysis, it may have to be considered. It is particularly important to consider inflation when the price of an input or a product is the variable of interest in a sensitivity analysis. As was discussed in Chapter 4, price changes may be divided into changes due to inflation and relative price changes. A sensitivity analysis must differentiate between these two types of changes. A price increase due to inflation will generally affect the prices of all other items in a similar manner and certainly will not have the same effect as a relative price change that changes the price of only one item.

Where possible, it is probably best to deal with constant dollar prices in a sensitivity analysis. Then only the relative prices are shown, the sensitivity to relative price changes can be seen, and inflation is effectively eliminated from the sensitivity analysis (although inflation is still considered in the economic calculation, of course). Dollar prices will change over the analysis period, making a sensitivity analysis on dollar prices difficult. Where it is necessary to use dollar prices as the subject of a sensitivity analysis, it must be determined whether the price differences are due to inflation or relative price changes. Other prices in the analysis must be adjusted accordingly if the difference is due to inflation.

9.2 EVALUATION OF SENSITIVITY

Conducting a sensitivity analysis on inflation may be more difficult than would appear to be the case. It is not possible to simply change the value of this one parameter and proceed with the calculations in most problems. The inflation rate determines the relationship between real and apparent time value, between dollars and constant dollars, and between real and apparent rates of return. As long as inflation is considered constant, these relationships stay constant, and consistent comparisons can be made by using either dollars or constant dollars, apparent or real time value. Changing the value of inflation means that these relationships are no longer fixed, and incorrect comparisons will be made if one is not careful.

Three measures of economic performance have been used thus far: present value, annual cash flow, and rate of return. The first two may be expressed either in dollars or constant dollars and the last one, in real or apparent terms. Calculating either of the first two requires that a time value also be specified in either real or apparent terms. This last requirement is particularly bothersome since if an apparent time value of money is set, the real time value must vary with the inflation rate, and if the real time value is set, then the apparent time value varies with inflation. A basic question must be asked as to which is the more fundamental and more constant parameter, real time value or apparent time value. It was argued in Chapter 3 that the real time value is approximately constant and that the apparent time value varies to reflect the inflation rate. On this basis it is best to consider the real time value as the constant parameter when making a sensitivity analysis for inflation. Unfortunately, most corporate economic analysis procedures are designed for using a fixed time value of money, which is approximately the same thing that is referred to as the apparent time value here. Some resistance may be encountered in making this parameter a function of inflation.

This problem of deciding whether to let real or apparent time value be constant may be partially circumvented by using a rate of return comparison where possible. Since no time value needs to be specified to calculate a rate of return, the calculation may proceed without addressing this problem. It is still important to realize the difference between real and apparent rate of return and that as inflation changes, so does the relationship between these two parameters. A formal decision still requires a time value to be specified; however, the purpose of a sensitivity analysis is more to provide additional information rather than to make yes/no decisions. A sensitivity analysis expressed in real rate of return may be sufficient for this purpose even though a time value is not specified. At some point, the decision as to what the correct time value is must be addressed. This topic is discussed further in Chapter 11.

Given the argument that the real time value is the proper variable to set when studying the effects of inflation on economics, it follows that evaluations of sensitivity to inflation must be made in constant dollar or real terms. Dollar revenues and costs increase with inflation and will increase the dollar present value, dollar annual cash flow, or the apparent time value accordingly. However, these measures show a false sensitivity to inflation since they increase as inflation increases even though everything stays constant in value terms. An investment that is insensitive to inflation is one whose value does not change in constant dollar terms when the inflation rate changes. The dollar or apparent measures of such investments vary with inflation according to Equations 5.1, 6.8, or 7.6.

9.3   FIXED TERMS

Many engineering economic analyses are sensitive to the inflation rate. However, those analyses that involved fixed dollars terms can be particularly sensitive to inflation. Fixed terms may result from contractual agreements or any other arrangement that sets prices or cash flows over a period of time. If incomes or the prices of products are fixed in dollar terms and costs are not, expenses can rise as a result of inflation, and the fixed incomes may no longer be adequate to cover the cost of production. Such situations may lead not only to investments with poor returns if inflation is higher than expected, but also to unexpected cash flow problems. Of course, the opposite situation arises when inflation is less than expected; the investment will yield better than expected. If expenses or prices of inputs are fixed, the reverse is true. Inflation rates less than expected can result in poor return from an investment and creates the potential for cash flow problems.

It is somewhat ironic that analyses involving fixed terms have this problem. Such arrangements are often made to eliminate uncertainty or risk. Since the terms are set, it is felt that everyone concerned knows exactly what to expect, and there is nothing to be uncertain about. However, these arrangements set the dollar cash flows or dollar prices. The actual value of the resulting cash flows then depends totally on the inflation rate. The sensitivity to inflation becomes particularly acute when long periods of time are also involved. A fixed term agreement is not greatly affected by one or two years of inflation unless the rate is extremely high. Several years of moderate inflation can result in these fixed dollar cash flows being worth only a fraction of the value of the same dollar cash flows now. Such arrangements pose substantial risks to both parties of such a transaction. The amount of risk is proportional to the part of the total cash flows that comes from fixed term agreements. Anytime an analysis involves a significant portion of fixed terms, it is very important that the sensitivity to inflation be calculated.

High grade corporate bonds, certificates of deposit, savings bonds, and other similar itmes are usually considered to be risk free investments. This conclusion comes only from the fact that there is almost zero probability that the dollar payments called for will not be made. These investments in fact are very uncertain when there is uncertainty in the inflation rate since they consist totally of fixed dollar cash flows. Such investments may actually be more risky than investments with a potential for failure but that have cash flows that increase with inflation if they do not fail. The uncertainty of the so-called safe investments can be estimated with Equation 8.13. The sensitivity of such an investment, one that pays 15% annual interest, is shown in Figure

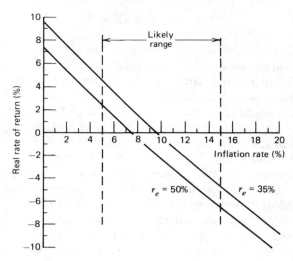

Figure 9.1 Sensitivity to inflation of an investment paying 15% interest.

9.1. Two curves are shown, one for an effective income tax rate of 35% and one for an effective income tax rate of 50%. The inflation rate for the next few years probably cannot be predicted more accurately than shown by the likely range on the graph. Within this range, there is considerable uncertainty as to the actual real rate of return that this investment would yield. The inflation rate for longer periods of time could be just about anywhere on the graph, resulting in a great deal of uncertainty for this investment.

9.4 COMPOUND RISKS FROM INFLATION

The potential consequences of changes in the inflation rate may be under-estimated by looking at the sensitivity of individual investments to inflation. Other uncertain variables usually affect only a few investments the same way. One investment may be highly dependent on the relative price of electricity, another on the demand for a particular electronics product, another on the amount of a chemical that must be processed, and so on. The variations in these factors are fairly independent, and the random variations often tend to offset each other. One investment will do better than expected and another, poorer than expected. The same inflation rate will usually apply to all the investments of a single investor or company. Even though there are many individual investments dealing with a number of products and inputs, it is quite possible that most of the investments are affected the same way by inflation. Rather than having a few losing investments, the investor or company may face serious hardships as the result of an unexpected major shift in the inflation rate. Those who seek diversity of investments to minimize risks should also ensure that there is a diversity of reaction to inflation. For example, a company that is locked into a number of large long-term projects where revenues were fixed would have difficulty surviving an extended period of unexpectedly high inflation.

9.5 EXAMPLES

Example 9.1

A heat exchanger manufacturer is considering supplying the evaporators for the heat pumps of an original equipment manufacturer. The heat exchanger company will enter into a contract with the heat pump company to supply heat exchangers if the manufacturer will guarantee a price of $120 per unit for the next three years. In turn, the heat pump company will guarantee to purchase at least 1200 units per year for the three year period. The manufacturer of the heat exchanger will require the purchase of a special machine to make and install fins on the heat exchanger. This machine will cost $250,000. The heat exchangers would require $15 worth of materials at today's prices and also about $20 worth of labor and other expenses to produce. The special machine has an estimated salvage value of $30,000. Assuming that only the number of units specified in the contract are sold, how does this project depend on the inflation rate?

The problem statement does not say so, but presumably all the costs stay constant in value terms, except, of course, the price of the end product. Therefore, the constant dollar cash flows for production of the evaporators are not affected by inflation. These cash flows are as follows where the present is used as the reference time:

Year	Constant Dollar Cash Flow for Production Cost
0	250,000
1	$(15 + 20) \cdot 1200 = 42,000$
2	$(15 + 20) \cdot 1200 = 42,000$
3	$(15 + 20) \cdot 1200 - 30,000 = 12,000$

This calculation also assumes that the salvage value stays the same in constant dollars. The income from the sale of the evaporators will decrease in constant dollar terms due to inflation. The constant dollar income will be as follows:

Year	Constant Dollar Income
0	0
1	$1200 \cdot \$120 \cdot \dfrac{1}{1+f} \quad \dfrac{\$}{\$}$
2	$1200 \cdot \$120 \cdot \dfrac{1}{(1+f)^2} \quad \dfrac{\$}{\$}$
3	$1200 \cdot \$120 \cdot \dfrac{1}{(1+f)^3} \quad \dfrac{\$}{\$}$

The resulting constant dollar net cash flows for various inflation rates are as follows:

Year	0% ($)	5% ($)	10% ($)	15% ($)	20% ($)
0	−250,000	−250,000	−250,000	−250,000	−250,000
1	+102,000	95,100	88,900	83,200	78,000
2	+102,000	88,600	77,000	66,900	58,000
3	+132,000	112,400	96,200	82,700	71,300

The real rate of return for each set of cash flows can now be calculated. The resulting real rates of return are as follows and are also shown graphically in Figure 9.2:

Inflation rate (%):	0	5	10	15	20
Real rate of return (%):	15.6	8.8	2.5	−3.4	−8.9

It is seen that this project is fairly sensitive to inflation. If the inflation rate is high, the project could be a losing proposition, but if the inflation rate should be moderate or low, it will make a fair to good return.

Example 9.2

A large firm utilizes a number of small computer systems. They presently employ a full time technician to maintain and repair these systems. The cost

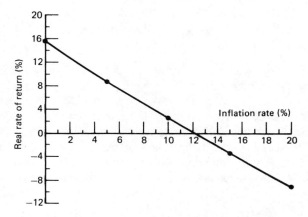

Figure 9.2   Effect of inflation on return in Example 9.1.

of the technician is presently $35,000 per year including fringe benefits and overhead expenses. Salaries and related expenses are expected to increase at about the same rate as inflation. A computer service company has offered to contract to service the computer systems for a flat rate of $40,000 per year for the next four years. The company estimates its real time value to be about 4%. How is the value of this contract affected by inflation?

The cost for the technician is expected to stay the same in constant dollar terms. The cost of the service contract will decrease in constant dollar terms if there is inflation. The present value of the cost of the contract is

$$PV_c = \left(\frac{P}{A}, i_a, 4\right) \cdot \$40,000$$

where

$$i_a = 1.04 \cdot [1 + f] - 1.0$$

This present value is in dollar terms, but if we let the reference year for constant dollars be the present, it will also be in constant dollars. The present value of the cost of the technician is

$$PV_t = \left(\frac{P}{A}, 4\%, 4\right) \cdot \$35,000$$

This present value is in constant dollars. The net present value of the contract in constant dollars is then

$$PV_n = \left(\frac{P}{A}, 4\%, 4\right) \cdot \$35{,}000 - \left(\frac{P}{A}, i_a, 4\right) \cdot \$40{,}000$$

The present value as a function of the inflation rate is shown in Figure 9.3. It is seen that the service contract is not profitable unless there is at least a moderate amount of inflation. However, it is profitable at recent inflation rates. It might also be noted that either dollar terms or constant dollar terms could have been used here since the present value is the same in either case. However, it is very important that the apparent time value of money was made dependent on inflation while the real time value was kept constant. If this measure had not been taken, completely different results would have occurred.

Example 9.3

A manufacturing plant is modernizing some of its production facilities. One particular machine could be fully automated for an additional \$113,000. Full automation will not save much labor in this case but will improve product quality and consistency. Presently 10% of the components that pass through this operation are not satisfactory. It is expected that by automating the machine less than 1% will be unsatisfactory. The automated machine should cost no more to operate than the conventional machine. Both have useful lives of about five years. The value of each lost component is \$35 at today's prices. The throughput for this plant varies with demand but is typically 10,000 units per year.

The automation expense can be considered to be depreciated by the straight-line method over five years with zero salvage value. The equipment also has zero economic salvage value. The company has an effective tax rate of 49%. The future price of the products of the plant are uncertain, but they

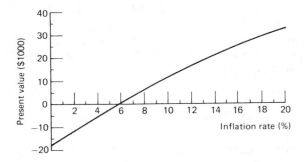

Figure 9.3 Effect of inflation on the value of the contract in Example 9.2.

usually follow overall price levels fairly closely. How will inflation affect the return on this investment before and after tax?

A complete sensitivity analysis would examine all the parameters and their effect on the return of this investment, particularly the effect of the through-put. All that was asked for here was to determine the effect of inflation before and after tax. Let us look at the before tax situation first.

The dollar revenue for this investment depends on the inflation rate. At today's prices there is an annual savings of $31,500 (900 fewer rejects at $35 each) with a throughput of 10,000 per year. The dollar savings will increase as the price of the component increases due to inflation. The following list shows the dollar cash flows for several inflation rates:

<table>
<tr><td colspan="5" align="center">Inflation Rate
($)</td></tr>
<tr><td>Year</td><td>0%</td><td>5%</td><td>10%</td><td>15%</td></tr>
<tr><td>0</td><td>−113,000</td><td>−113,000</td><td>−113,000</td><td>−113,000</td></tr>
<tr><td>1</td><td>31,500</td><td>33,075</td><td>34,650</td><td>36,225</td></tr>
<tr><td>2</td><td>31,500</td><td>34,729</td><td>38,115</td><td>41,659</td></tr>
<tr><td>3</td><td>31,500</td><td>36,465</td><td>41,927</td><td>47,908</td></tr>
<tr><td>4</td><td>31,500</td><td>38,288</td><td>46,119</td><td>55,094</td></tr>
<tr><td>5</td><td>31,500</td><td>40,203</td><td>50,731</td><td>63,358</td></tr>
</table>

The apparent rate of return (ROR) may be caluclated from these cash flows and then converted to a real rate of return. The results are as follows:

Inflation rate (%):	0	5	10	15
Apparent ROR (%):	12.2	17.8	23.4	29.0
Real ROR (%):	12.2	12.2	12.2	12.2

The before tax real rate of return is seen to be totally unaffected by inflation, which is not totally unexpected since savings were assumed to increase exactly at the inflation rate. It should also be noted that the apparent rate of return shows a strong dependency on the inflation. It is very important that this result not be interpreted as meaning that the investment improves with inflation.

Now let us turn to the after tax situation. The savings from the investment will be taxed at the rate of 49%. However, the depreciation allowance will decrease this tax somewhat. The depreciation each year will be

$$\frac{\$113,000}{5} = \$22,600$$

This dollar depreciation allowance will not be affected by inflation. The taxable income due to the investment will be the dollar savings less this depreciation allowance. The after-tax cash flows for various inflation rates are tabulated in the following list:

Year	Before-Tax Cash Flow ($)	Taxable Income ($)	Tax ($)	After-Tax Cash Flow ($)
		$f = 0\%$		
0	−113,000	0	0	−113,000
1	31,500	8,900	4,361	27,139
2	31,500	8,900	4,361	27,139
3	31,500	8,900	4,361	27,139
4	31,500	8,900	4,361	27,139
5	31,500	8,900	4,361	27,139
		$f = 5\%$		
0	−113,000	0	0	−113,000
1	33,075	10,475	5,133	27,942
2	34,729	12,129	5,943	28,786
3	36,465	13,865	6,794	29,671
4	38,288	15,688	7,687	30,601
5	40,203	17,603	8,625	31,578
		$f = 10\%$		
0	−113,000	0	0	−113,000
1	34,650	12,050	5,905	28,746
2	38,115	15,515	7,602	30,513
3	41,927	19,327	9,470	32,457
4	46,119	23,519	11,524	34,595
5	50,731	28,131	13,784	36,947
		$f = 15\%$		
0	−113,000	0	0	−113,000
1	36,225	$13,625	$6,676	29,549
2	41,659	19,059	9,339	32,320
3	47,908	25,308	12,401	35,507
4	55,094	32,494	15,922	39,172
5	63,358	40,758	19,971	43,387

The apparent rates of return for each set of after tax dollar cash flows can now be calculated and used to determine the real rate of return in each case. The results are as follows:

Inflation rate (%):	0	5	10	15
Apparent ROR (%):	6.4	9.7	13.0	16.5
Real ROR (%):	6.4	4.5	2.8	1.3

There is now some sensitivity to the inflation rate. The real rate of return decreases as inflation increases. This result is due totally to the effects of taxation since the before tax real rate of return was unaffected by inflation. The apparent rate of return again gives a false indication of sensitivity, showing that the investment improves with inflation. This particular investment is relatively insensitive to inflation as compared to the two previous examples. The two previous examples involved fixed terms and were directly affected by inflation. The investment in this example is only indirectly affected by inflation through taxation. The sensitivity is still sufficiently large to affect the decision, however.

Example 9.4

A farmer is considering on-the-farm production of alcohol as a motor fuel for his tractors. He has analyzed the cost of setting up and operating the fuel production facilities and has come up with the following results:

Initial cost of still, brewing vats, and storage tanks: $7000.

Cost to modify tractors for alcohol fuel: $250 each for three tractors.

Cost of grain: $4.50/bu. Each bushel yields about 3 gal of fuel, and there is also a residue that is worth $3.50 as animal feed for each bushel of grain consumed.

Labor expense: The farmer considers his labor worth $7/h. The still must be attended continuously when in operation and is fueled by wood cut locally that has no other economic value. It takes 2 h to cut and haul the wood for 1 h of operation. Also, 16 h of labor is required for mashing and setup for each 500 gal batch.

Throughput: The still has a capacity of 50 gal/h, and the farmer uses 500 gal of fuel each month on the average. This fuel replaces 450 gal of gasoline.

Life: The still and other components should last about 10 years with relatively little maintenance other than routine cleaning.

The economic benefit of this operation would derive from eliminating the use of gasoline in the tractors. Thus the economic viability of the operation depends on the price that must be paid for gasoline. The farmer would like to know how expensive gasoline must be before this on-the-farm production is worthwhile and how good the investment is at various prices.

It is very dangerous to attempt to come up with a single gasoline price for which the fuel production becomes economic. As was discussed earlier, the prices of nearly everything else will increase when the price of the fuel increases as a result of inflation. However, it is possible to determine the price in today's dollars for which the operation become feasible. This calculation is equivalent to holding everything else constant and letting the price of gasoline increase. The results must then be interpreted properly to be meaningful.

No information is given about expected price changes of other factors. Since we want to study gasoline prices only, it is reasonable to assume that all other prices stay the same in constant dollars so the analysis can proceed using the cost information given. The cost per 500 gallon batch is

$$500 \text{ gal} \cdot \frac{1 \text{ bu}}{3 \text{ gal}} \ [\$4.50/\text{bu} - \$3.50/\text{bu}] \\ + 46 \text{ h} \cdot \$7/\text{h} = \$489$$

An output of twelve batches per year is required, resulting in a total annual operating cost of $\bar{\$}5864$. The total initial cost for the fuel production equipment and the conversion cost is $\bar{\$}7750$. The gasoline savings each year is

$$12 \cdot 450 \text{ gal} \cdot p^c_{\text{gasoline}}$$

The net present value of the investment is then

$$PV = -\bar{\$}7750 + \left(\frac{P}{A}, i_r, 10\right) \cdot [5400 \cdot p^c_{\text{gasoline}} - \$5864]$$

The real rate of return of the investment as a function of the gasoline price can be found by setting this present value to zero. The results are shown in Figure 9.4.

The real rate of return of the investment is seen to be extremely sensitive to the price of gasoline. This result is understandable since the annual cash flows are about as large as the initial cost and the net annual cash flows is the difference between two large numbers. No required rate of return is given, but it is seen that the actual real rate of return is reasonably good if the price of gasoline is $1.25 to $1.30 per gallon and greater.

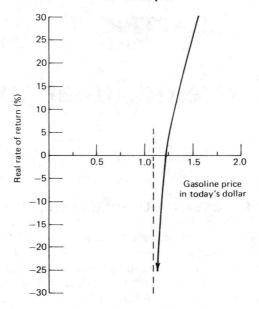

Figure 9.4 Effect of gasoline price on the rate of return in Example 9.4.

In this particular sensitivity analysis, the important thing to remember about inflation is that it must be removed from the analysis. The results must now be properly interpreted. Just because the dollar price of gasoline exceeds $1.30/gal does not mean that the investment is worthwhile. It is only if the price exceeds $1.30 in constant dollars where today is the reference time that the investment becomes worthwhile. For example, if there is 10% annual inflation for the next five years, the price should be

$$\$1.30 \cdot [1.10]^5 \frac{\$}{\$} = \$2.09/\text{gal}$$

for the investment to be economic, assuming that all other prices increase at the same rate as inflation. It is particularly important, with analyses like this, to realize that price increases due to inflation have little or no effect on the economics. It is only when the prices change in relative terms that the economics change.

CHAPTER TEN

Extreme Inflation

10.1   LOSS OF A USABLE STANDARD

The ideal dollar* for economic analysis would have a constant buying power. Then a given cash flow would represent the same amount of value whenever it occurred. Such a dollar could be used as a standard of value for economic calculations. However, the value of the dollar does change, and hence it is a poor standard for measurement of economic value. The creation of an artificial constant dollar as has been described in preceding chapters is an attempt to create the ideal standard of value. Inflation has been treated simply as an imperfection in the dollar value measure, for which corrections must be made.

As long as inflation rates are very low, no such corrections are really necessary and the dollar may be treated as an unchanging standard for measurement. When the value of the dollar varies significantly, such as it has in the last decade or two in the United States, it is necessary to make corrections as have been described in the preceding chapters. The greater the inflation rate, the more important it is to make corrections for the changing value of the dollar. As long as the inflation rate is not too large, say, less than 15–20%, it is fairly practical to make the corrections described here. These inflation rates may seem rather high to many people, but on a world standard they are fairly moderate.

Unfortunately, when inflation rates become large and it is even more important to account for the effects of inflation, it also becomes more difficult to do so. The problem does not arise from a breakdown of the theory. The theory is generally valid even for high inflation rates. The problem comes from the inability to apply the theory. To correct the dollar for changing value, a fairly reasonable prediction of future inflation rates is necessary. When the inflation rate is moderate, say about 10%, economists are still able

---

*The reader is reminded that the dollar is used here to represent currency in general and not just the U.S. dollar.

to predict the inflation rate for future years within $\pm$ 5 percentage points. Although this degree of uncertainty is sufficient to cause significant problems for many economic analyses, it is an uncertainty that can usually be dealt with fairly well. Forces in the economy that cause inflation rates of 20–30% and greater usually tend to be fairly unstable and result in erratic inflation rates. Inflation may be 30% one year, 100% the next, 50% the next, and so on. Such conditions make it essentially impossible to make corrections for the value of a dollar. There is no way that accurate predictions can be made in these circumstances.

The result of these extreme inflation rates is that the dollar can no longer be used or corrected and used as a standard for the measure of value. It is even more important to be concerned about the effects of inflation in this case than with moderate inflation rates. However, the formal mathematical calculations are of little value. These situations do not lend themselves to a simple theoretical analysis or rational decision making. Individual judgment becomes more important. Economic analysis and decisions must be based as much as possible on consideration of physical exchanges, such as barrels of oil produced, the amount of grain harvested, and the amount of steel required. Hopefully, the relative prices for the various products will not change too greatly, and profitable ventures at one point in time will still be profitable after a several hundred percent increase in prices. If contracts or other fixed term arrangements are involved, analysis becomes very difficult and the results become very uncertain. Removing some of the uncertainty is the fact that, with extreme inflation, fixed-term cash flows will usually have very little value after just a few years because of the rapid decline in value of the dollar.

TABLE 10.1 Inflation in Latin America

Country	Inflation Period	Price Index Ratio[a]	Equivalent Annual Inflation Rate for Period[b] (%)
Argentina[1]	1947–1960	21.5	27
	1960–1974	29.5	27
Brazil[1]	1947–1960	8.0	16
	1960–1974	77.0	36
Chile[1]	1947–1960	34.5	31
	1960–1971	11.6	25
Uruguay[2]	1949–1960	3.1	11
	1960–1970	52.5	49

[a]Ratio of the price index at the end of the inflation period to the price index at the beginning of the period.
[b]The annual inflation rate that would give the same price index change for the same time period.

Incidences of extreme inflation are not rare. In fact, extreme inflation is more the rule than the exception in many parts of the world. Latin America is a typical example. Table 10.1 shows the inflation rates that have been experienced in a number of these countries. These countries often experience erratic as well as high inflation rates. As an example, the annual inflation rates in Uruguay are shown in Table 10.2. The future value of a currency under such conditions is essentially impossible to predict.

10.2   HYPERINFLATION

Although inflation rates like those shown in Tables 10.1 and 10.2 certainly are not good and are difficult to deal with when trying to assess the economics of engineering and other alternatives, things can get much worse. Sometimes conditions arise that cause explosive growth in prices. Prices may increase by a factor of 10, 100, or even more in a single year. Such inflation is referred to as hyperinflation. Hyperinflation is the result of a complete breakdown in the value of a currency, and it no longer has any usefulness as a standard of value. Remember that the value of a paper currency exists only when people

TABLE 10.2   Inflation in Uruguay[2]

Year	Inflation Rate (%)
1955	8
1956	7
1957	15
1958	17
1959	40
1960	39
1961	22
1962	11
1963	21
1964	43
1965	56
1966	73
1967	89
1968	125
1969	21
1970	16

accept it as having a value. When the confidence in a currency is lost, its value can decline very rapidly.

It is beyond the scope of this discussion to thoroughly explore the causes and nature of hyperinflation. However, it is probably correct to say that hyperinflation has almost always resulted from a government trying to purchase far more goods and services than it has revenue to pay for. The difference is made up by creating more money in one form or another. This unwarranted creation of money for which no goods or services are exchanged can quickly destroy a currency and result in hyperinflation. The process is accelerated when the government attempts to continue buying. Money is less valuable because of the inflation, and it must create even more money to maintain its level of purchasing. A vicious spiral to runaway inflation results.

No amount of adjustment, correction, or calculation can make a currency of any use for measuring long-term value when hyperinflation occurs. Any economic analysis must rely entirely on physical measures. Further complicating any economic analysis is the social and political unrest that generally accompanies hyperinflation. The only predictable effect of hyperinflation is to essentially eliminate all dollar debts, all paper investments, and all fixed term agreements. When the value of a dollar declines by a factor of 10, 100, 1000, or more in a year's time, the value of anything specified in dollar terms becomes negligible very quickly, sometimes in just a few days. Debts, bonds, life insurance, savings, and other such items purchased before the hyperinflation begins are totally lost in the inflationary spiral. This is why many people accumulated gold, silver, and other items with tangible value when they feel inflation rates are likely to become large.

There have been many cases of hyperinflation throughout history. Some of the better known and better documented cases are shown in Table 10.3. Hyperinflation is not as uncommon as one might think. As mentioned earlier,

TABLE 10.3   Incidences of Hyperinflation

Country	Inflation Period	Price Index Ratio[a]	Equivalent Annual Inflation Rate for Period[a] (%)
China[3]	August 1948–April 1949		
(Shanghai)		1.04×10^6	1.67×10^9
(Chungking)		5.48×10^4	4.86×10^7
Germany[4]	July 1921–November 1923	5.24×10^{10}	3.93×10^6
Hungary[5]	July 1945–July 1946	3.81×10^{29}	2.02×10^{29}

[a] See Table 8.1 for explanation.

it usually occurs along with social and political upheaval. Hyperinflation has occurred twice in the United States: during the Revolutionary War and in the Confederacy during the Civil War.

REFERENCES

1. *Latin American Inflation*, Susan M. Wachter, Lexington Books, Heath, Lexington, MA, 1976, pp. 123–134.
2. *Chronic Inflation in Latin America*, Felipe Paxos, Praeger Publ., New York, 1972, p. 114.
3. *The Chinese Inflation 1937–1949*, Shun-Hsin Chou, Columbia University Press, New York, 1963, p. 20.
4. *The German Inflation of 1923*, Fritz K. Ringer, Ed., Oxford University Press, London, 1969, p. 80.
5. *Inflation*, Michael Jefferson, Andrew D. White, Thomas Mann, and Walt Rostow, John Calder Publ., 1977, p. 50.

CHAPTER ELEVEN

Variable Inflation Rates and Time Value

Time value, either real or apparent, and the inflation rate have thus far been treated as if they were constant over the entire analysis period for a given evaluation. These assumptions have been made in spite of the fact that inflation rates as reflected in the price indexes have varied substantially over the past 10–20 years. A review of the time value of money used by most any firm will show that it has varied over this time period also. One may then be inclined to ask what justification there is for using constant inflation rates and constant time values as has been done thus far. The primary justification is the uncertainty of future values of these parameters, particularly the inflation rate. There is so much uncertainty in the inflation rate for the future that it is seldom deemed worthwhile to worry about variations with time. Apparent time value tends to follow inflation as was shown in Chapter 3. Its uncertainty will then be comparable to that for inflation. Only the real time value seems to remain fairly constant with time. Once the assumptions of constant time values and inflation rates are made, economic analyses are simplified. Problems may be solved by one of two methods: constant dollars-real time value or dollars-apparent time value. Totally consistent results are obtained with the two methods. The assumptions of constant inflation and time values also allow a number of mathematical relationships to be developed that simplify economic analysis calculations.

The effect of changing the inflation rate was investigated in Chapter 9. However, the concept was different there. It was still assumed that the inflation rate stayed constant over the entire analysis period. The effect of changing this constant value was all that was examined. Here the focus of attention is the effect of an inflation rate that changes during the analysis period.

151

11.1 EXPRESSING INFLATION

If a constant inflation rate is not to be used, some method for expressing the more complex projection of inflation is required. The most straightforward approach is to simply project future values of the ratio

$$\frac{\bar{p}(t)}{\bar{p}(t_0)}$$

which is a proper description of inflation. The relationships between constant dollars and dollars were developed by using this ratio, and the projected ratio may be substituted directly into the appropriate equations. The term

$$[1 + f]^n$$

was substituted for the price ratio when the inflation rate was constant. This term is no longer meaningful when the inflation rate varies.

Most people are used to expressing inflation in the form of a rate and are more comfortable with this form even when the rate does not stay constant. There is no problem with expressing future inflation in the form of rate that varies with time. For example, the inflation rate might be projected to be 15% next year, 12% the two following years, and 10% thereafter. This form of projection must be converted to a price ratio form before calculations can be made. The inflation rate for a single time period was defined as

$$f_1 = \frac{\bar{p}(t_1) - \bar{p}(t_0)}{\bar{p}(t_0)} \qquad (11.1)$$

or

$$1 + f_1 = \frac{\bar{p}(t_1)}{\bar{p}(t_0)} \qquad (11.2)$$

where t_1 and t_0 are one time period Δt apart. This relationship may be expanded to multiple time periods with different inflation rates:

$$\frac{\bar{p}(t)}{\bar{p}(t_0)} = [1 + f_1] \cdot [1 + f_2] \cdot \cdots [1 + f_j] \qquad (11.3)$$

where f_1 is the inflation rate from t_0 to t_1, f_2 is the inflation rate from t_1 to t_2, and so on, and

$$j = \frac{t - t_0}{\Delta t}$$

Either method of projecting inflation results in an expression of the price as a function of time. It is also seen that the expression

$$\frac{\bar{p}(t)}{\bar{p}(t_0)} = [1 + f]^n$$

used with constant inflation rates is only a special case of Equation 11.3. The projected price ratio may be substituted directly into many of the relationships derived previously. However, this simple calculation for price ratio by itself is not sufficient to allow economic analysis to be made when the inflation rate varies. Some more difficult questions about time value must be resolved first.

11.2   CONSISTENCY BETWEEN DOLLAR AND VALUE METHODS

The previously derived relationship between the inflation rate and real and apparent time value

$$1 + i_r = \frac{1 + i_a}{1 + f}$$

was a result of requiring equivalence calculations made in constant dollars and the real time value to be consistent with equivalence calculation made in dollars and the apparent time value. All three variables were assumed constant in this derivation. It is obvious that if the inflation rate is allowed to vary, the real time value and the apparent time value cannot both remain constant and maintain this relationship. It is further shown in Appendix 11.1 that if both the apparent time value and the real time value are constant over a period of time, the only case where consistency can be guaranteed between the two methods for making equivalency calculations is when the relationship

$$\frac{\bar{p}(t)}{\bar{p}(t_0)} = [1 + f]^{[(t - t_0)/\Delta t]}$$

is valid during this time period. This result is analogous to saying that the inflation rate must be constant if both real and apparent time value are to be constant.

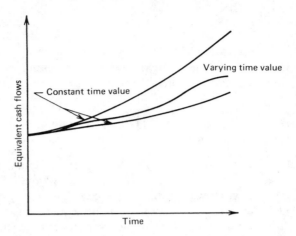

Figure 11.1   Constant and varying time values of money.

Once the consistency between comparison methods breaks down, the ability to compare and evaluate alternatives also breaks down. It is shown in Appendix 11.2 that all three of the methods that were used before—present value, annual cash flow, and rate of return—may yield inconsistent results when the real and apparent time values are held constant and inflation is allowed to vary.

It does little good to point out all the problems when a varying inflation rate exists unless some improvements can be made. Recall from Chapter 3 that in simple theoretical terms the time value of money is determined in the marketplace. The general concept is that, through the market place, a given amount of funds at one point in time can be exchanged for a different amount of funds at a different point in time. The relative difference in funds from one time period to the next is the time value:

$$i = \frac{Y(t_2) - Y(t_1)}{Y(t_1)} \qquad (11.4)$$

where i is the time value for the period from t_1 to t_2. There is nothing in this simple theory that implies the time value will be the same from one period to the next. This assumption was made only to simplify economic calculations. Figure 11.1 shows equivalent value functions that reflect constant time value and varying time value. There is no reason why the varying situation cannot be used in development of the economic theory.

The development of this theory in Chapter 3 was based on an ideal market where it was assumed, among other things, that everybody had perfect knowledge about investments and related factors. The perfect knowledge would include the value of the inflation rate in future years. The equivalence curve established by the market then should reflect the inflation that will occur in future years. Since perfect foreknowledge exists, the relative equivalent amounts between two specific points in time will not change in the future. Factors such as inflation that will affect the time value are already reflected in the equivalent value function.

Assuming the market functions in dollar terms, equivalency calculations may be made by the relationship

$$Y^d(t_1) = \frac{E^d(t_1)}{E^d(t_2)} \cdot Y^d(t_2) \qquad (11.5)$$

where $Y^d(t_1)$ and $Y^d(t_2)$ are equivalent amonts, at times t_1 and t_2, respectively, and $E^d(t_1)$ and $E^d(t_2)$ are derived from the equivalent value function as shown in Figure 11.2. In this situation of perfect knowledge, the values of the price ratio

$$\frac{\bar{p}(t)}{\bar{p}(t_0)}$$

will also be known for each point in time. Dollar cash flows may still be converted to constant dollar cash flows by applying this ratio:

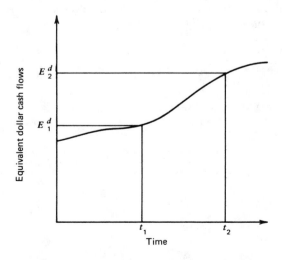

Figure 11.2 Determination of Equivalent cash flows.

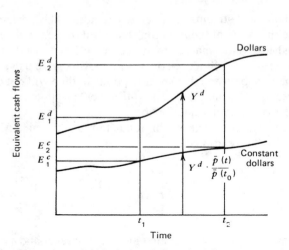

Figure 11.3 Comparison of equivalent cash flows in dollars and in constant dollars.

$$Y^c(t) = Y^d(t) \cdot \frac{\bar{p}(t_0)}{\bar{p}(t)} \qquad (11.6)$$

This calculation may be applied to the dollar time value function in Figure 11.2 to create equivalent values in constant dollar form as is shown in Figure 11.3. The constant dollar time value relationship may also be derived mathematically. Equation 11.6 can be used to express Equation 11.5 in constant dollar form:

$$Y^c(t_1) \cdot \frac{\bar{p}(t_1)}{\bar{p}(t_0)} = \frac{E^d(t_1)}{E^d(t_2)} \cdot Y^c(t_2) \cdot \frac{\bar{p}(t_2)}{\bar{p}(t_0)} \qquad (11.7)$$

where $Y^c(t_1)$ and $Y^c(t_2)$ are now equivalent constant dollar cash flows at times t_1 and t_2, respectively. Since

$$\frac{\bar{p}(t_2)}{\bar{p}(t_1)} = \frac{\bar{p}(t_2)/p(t_0)}{\bar{p}(t_1)/\bar{p}(t_0)} \qquad (11.8)$$

Equation 11.7 may be reduced to

$$Y^c(t_1) = \frac{E^d(t_1)}{E^d(t_2)} \cdot \frac{\bar{p}(t_2)}{\bar{p}(t_1)} \cdot Y^c(t_2) \qquad (11.9)$$

Equivalent value ratios in constant dollar form may be defined by

$$Y^c(t_1) = \frac{E^c(t_1)}{E^c(t_2)} \cdot Y^c(t_2) \qquad (11.10)$$

where E^c is the constant dollar analogy of E^d. If Equations 11.9 and 11.10 are compared, it is seen that

$$\frac{E^c(t_1)}{E^c(t_2)} = \frac{\bar{p}(t_2)}{\bar{p}(t_1)} \cdot \frac{E^d(t_1)}{E^d(t_2)} \qquad (11.11)$$

Thus once the inflation function

$$\frac{\bar{p}(t)}{\bar{p}(t_0)}$$

is defined and the time value function

$$\frac{E(t)}{E(t_0)}$$

is defined, where the latter may be defined in either dollars or constant dollars, then equivalency calculations may be made in either dollar form or constant dollar form. Equations 11.5, 11.10, and 11.11 are the general relationships for equivalency calculations. Previously derived relationships are simplifications of these expressions and are valid only when the inflation rate and time values are constant.

11.3   CALCULATIONS WITH VARIABLE INFLATION RATES AND TIME VALUE

Arguments were made in Chapter 5 to show that the net present value is the proper economic comparison for alternatives. This argument was not based on any assumptions about constant inflation rates or time value and should be equally valid in the more general area of variable parameters presented here. The calculations may be somewhat more difficult than with constant parameters since interest factors can no longer be used. The present value for each cash flow must be calculated individually using Equations 11.5 and 11.10.

The annual cash flow method may also still be used, but the calculations are more involved. Recall that the annual cash flow was shown to be a valid comparison by the relationship

$$ACF = \left(\frac{A}{P}, i, n\right) \cdot PV$$

Two annual cash flows could be compared since they are related to the respective present values by the term

$$\left(\frac{A}{P}, i, n\right)$$

which is the same for each alternative as long as the same values for i and n are used for all alternatives. Thus a theoretically valid comparison may be made for any value of i and n. However, the resulting annual cash flow has no meaning. It is not equivalent to the present value; it is just proportional to it.

A meaningful annualization occurs only when what is referred to as an *equivalent uniform annual cash flow* results, in either constant dollars or dollars. The simple interest factor relationship cannot perform this transformation. The present value of a uniform series of cash flows can be found by

$$PV(t) = \sum_{j=1}^{n} ACF \cdot \frac{E(t)}{E(t_j)} \qquad (11.12)$$

where $E(t)$ may be either in dollars or constant dollars and ACF must be in the same terms. The annual cash flow for a given present value is then

$$ACF = \frac{PV(t)}{\sum\limits_{j=1}^{n} \cdot E(t)/E(t_j)} \qquad (11.13)$$

where n is the length of the analysis period. Since the term in the denominator is the same for all alternatives, the annual cash flow calculated in this manner is also proportional to the present value as well as being equivalent to it and may be used for economic comparisons.

It was noted in Chapter 6 that one of the most common applications of the annual cash flow calculation is for alternatives with unequal lives. Annualization over one life cycle is all that is necessary as long as cash flows recur from one life cycle to the next and the time value remains constant. When the time value does not remain constant, the annualization factor

$$\sum_{j=1}^{n} \frac{E(t)}{E(t_j)}$$

is not likely to remain constant from the time period for one life cycle to the time period for the next. Thus the annual cash flow for a single life cycle may not be the same for following life cycles even when the cash flows recur exactly. The more general method of comparing multiple life cycles may be necessary when the time value varies from one life cycle to the next.

The rate of return calculation presents greater, apparently insurmountable, problems when the time value is not constant. A rate of return may be calculated as before. This rate of return is the time value that results in a zero present value and is constant over the life of an investment. The difficulty comes in evaluating the result. When time value is constant, it is sufficient to show that

$$R > i$$

to establish the desirability for most investments.* No simple criterion can be established to evaluate the rate of return when the time value is not constant. This difficulty rules out a theoretically sound evaluation by the rate of return method. However, it still may be a useful tool for less rigorous evaluations.

It is possible that one form of the time value will remain constant, most likely the real time value, whereas the inflation rate and the other form of time value vary with time. The more complex calculations presented here are not necessary in such cases. As long as one form of time value remains constant, calculations may be made in the appropriate terms to use that time value, constant dollars for real time value, or dollars for apparent time value. It is not possible to switch back and forth between real and apparent time value, but calculations may be made as before with the time value that remains constant. In particular, valid rate of return comparisons may be made with the time value that is constant with time.

11.4 DISCUSSION

The preceding presentation leads to the conclusion that theoretically valid and consistent economic analyses may be made even if inflation and time value do not remain constant. The arguments are based on the theoretical

*This criterion applies to investments whose cash flow have only one sign reversal as shown in Figure 7.2.

situation of a perfect market with perfect knowledge. The real market is not perfect, and people certainly do not have perfect knowledge, particularly of the future. Thus some judgement must be employed in applying this, or any other, economic analysis theory to real problems. Two important points brought out in the theory should be empahsized. First, the assumption of perfect knowledge means that the equivalence function, or time value, reflected future conditions including the inflation rate. The inflation rate and the time values are not considered to be independent functions. Second, given the variation of the inflation rate with time and one form of the time value function, either the dollar form or constant dollar form, the other form of the time value function cannot be set independently but is derived from the first two. This requirement is not different than with the constant parameter assumption. Even though a perfect theoretical situation is not likely to exist for real world economic analyses, it is important to remember these two points when it is necessary to incorporate a varying inflation rate into an economic analysis.

It was shown in Chapter 3 that time value in dollar terms is strongly affected by the inflation rate. When a varying inflation rate is projected, it is particularly important to be concerned with how that inflation will affect the apparent time value. There is also good evidence that the real time value stays relatively constant over fairly long periods of time. These two considerations can be used to simplify the use of varying inflation rates. It is unlikely that sufficient information will exist to make a detailed assessment of time value variations in most cases. A reasonable approach is to let the real time value remain constant and let the time value in dollar terms vary according to the inflation rate by the relationships derived in this chapter. This approach has the advantage of letting one expression of time value be constant, and present value, annual cash flow, or rate of return calculations may be made in constant dollar form and applied in the same way as in previous chapters. The only difference is that conversion of dollars to constant dollars requires use of the ratio

$$\frac{\bar{p}(t)}{\bar{p}(t_0)}$$

and the expression

$$[1 + f]^n$$

cannot be used when inflation varies.

It was argued in Chapter 9 that real time value and dollar terms are the more meaningful expressions for comparing economics under changing conditions. That same argument applies here. When neither form of time value can be assumed constant, variations in the constant dollar expression of time value have more intuitive meaning. Investments in general will tend to earn more in real terms during times of economic prosperity than during times of economic stress. The inflation rate may or may not correlate to relative economic prosperity. It is reasonable to project a decrease in the real time value if declining economic conditions are projected and an increase if increasing prosperity is projected. However, it should be realized that there are many other factors that combine to determine either the real or apparent time value and that these simplistic concepts cannot be applied universally.

The theory and calculations discussed in this chapter presently are not included in most routine economic analyses. The assumptions of constant inflation rate and constant time value are usually adequate in view of the uncertainty of the future. However, when situations arise where one or more of these parameters varies, it is necessary to include the considerations presented in this chapter. Of particular importance is the situation where a constant apparent time value is specified for an economic analysis and at the same time a varying inflation rate is projected. Such specifications are almost certainly contradictory, and there is no way to resolve this difference unless one of the parameter specifications is changed.

11.5   EXAMPLES

Example 11.1

An oil pipeline company wants to temporarily reverse the direction of flow of a segment of a pipeline to make supplies of crude oil available to a refinery in short supply. The pipeline crosses a small ridge of mountains. Company engineers are trying to decide whether to install two small pumping stations or one large pumping station to raise the oil over the mountains on the side it previously ran down. Current cost estimates are as follows:

	Single Station	Two Stations
Installation cost:	$435,000	$585,000
Annual maintenance cost:	$150,000	$170,000
Annual electricity use:	5.67×10^6 kW · h	4.56×10^6 kW · h

The company plans to operate the pipeline in this mode for about five years. No salvage value is being assigned to the pumping stations. The cost for maintenance is expected to increase at about the same rate as inflation. The price of electricity is expected to increase in real terms at about 1% annually. The present price is 4.5¢/kW · h.

The company has established an apparent time value of money of 16% at the present rate of inflation of 11%. Company economists insist that economic analyses be based on the following projection of inflation:

Time from Present	Applicable Inflation Rate (%)
0–2 years	12
Thereafter	8

Given this information, evaluate the pumping station alternatives.

The 16% apparent time value probably will not stay constant if inflation behaves as projected. Probably the simplest solution in this example is to convert to a real time value and assume that it stays constant since no contrary information is given. The real time value is

$$i_r = \frac{1.16}{1.11} - 1.0 = 0.0450 = 4.5\%$$

The cash flows can now be stated in constant dollar terms and the problem solved just as any other problem would be. Note that once it is decided to work in constant dollars, the inflation projection does not enter into the calculation in this problem.

The electricity cost is projected to increase 1% annually in real terms, so these cash flows must be determined. The results are shown below using today as the reference time for the constant dollar calculations and using the year-end price convention.

Year	Electricity Price (S/kW · h)	Electricity Expense ($) Single Station	Electricity Expense ($) Two Stations
0	0.0450	—	—
1	0.0455	258,000	207,000
2	0.0459	260,000	209,000
3	0.0464	263,000	211,000
4	0.0468	266,000	214,000
5	0.0473	268,000	216,000

The comparison can be made by any one of the three methods. The rate of return method is used here. The net cash flows for each alternative and the incremental cash flow follow:

Year	Single Station ($)	Two Stations ($)	Difference ($)
0	435,000	585,000	−150,000
1	408,000	377,000	+31,000
2	410,000	379,000	+31,000
3	413,000	381,000	+32,000
4	416,000	384,000	+32,000
5	418,000	386,000	+32,000

The real rate of return is found using the net difference cash flow. Solving for the zero present value gives

$$i = 3\%, \quad PV = -\$5400$$
$$i = 2\%, \quad PV = -\$1110$$
$$i = 1\%, \quad PV = +\$3340$$
$$i = 1.75\%, \quad PV = \$17 \approx 0$$

The extra investment for the two station alternative earns a real rate of return of only 1.75%, which is less than the required 4.5%. The single station approach is the more economical.

The problem could have been solved in dollar terms. This approach would require setting up a function describing dollar value as a function of time. The results would be consistent with those obtained here, but a rate of return evaluation would not be valid. Example 11.2 demonstrates the solution of a similar problem in dollar terms.

Example 11.2

A manufacturing plant produces a lot of filings and other fine metal scrap. All the scrap is, at present, just mixed together and sold locally for recycling. The scrap metal would be worth considerably more if the different types of metal were not mixed together. The difficulty of keeping it all separated has prevented the company from taking advantage of this higher value. However, one particular operation is performed exclusively on brass. Because of the high value of brass, it has been proposed that a device to automatically collect and store the scrap produced by this process be added to the machine.

The collection device would cost about $35,000 to install and about $1000 per year to maintain and operate. Since the brass scrap is presently mixed in with other metals, it has little value, about $500 per year. If it is sold separately, it should be worth $7000 per year at today's prices. The machine to which the collection device is being added has an estimated life of eight years. Scrap metal prices tend to fluctuate fairly widely. It is anticipated, however, that these prices will increase at least as fast as overall inflation.

This company employs detailed economic forecasting for evaluation of their major investments and uses the same information for evaluating smaller investments such as this one. The inflation rate is currently 12%, and the time value of money established by the company translates into a real time value of 3.5%. They expect these values to hold the same for the next year. The real time value is then expected to increase by two percentage points for the following two years due to anticipated tight money, then drop down by about four percentage points for the following two years because of an anticipated depressed economy and then level off to 4% thereafter, a value that they feel is an accurate representation of the long-term real time value. The inflation rate is expected to stay at 12% for the next year, drop to 8% the following year, drop to 6% for the next year, and then increase by 1% per year until it levels off at 9%.

The first thing to do in assessing the economics of this investment with the use of all this nonconstant information is to construct a tabular list that converts the percentage rates to price ratios and time value ratios:

Year	Inflation Rate (%)	Real Time Value (%)	$\dfrac{\bar{p}(t)}{\bar{p}(0)}$	$\dfrac{E^c(t)}{E^c(0)}$	$\dfrac{E(t)}{E^d(0)}$
0	—	—	1.000	1.000	1.000
1	12	3.5	1.120	1.035	1.159
2	8	5.5	1.210	1.092	1.321
3	6	5.5	1.282	1.152	1.477
4	7	1.5	1.372	1.169	1.604
5	8	1.5	1.482	1.187	1.759
6	9	4.0	1.615	1.234	1.993
7	9	4.0	1.760	1.284	2.260
8	9	4.0	1.919	1.335	2.562

The economics may be evaluated in either dollars or constant dollars. Since the prices are expected to stay constant in constant dollar terms, this representation is the easiest. The net savings each year is

$$\$7000 - \$1000 - \$500 = \$5500$$

The present value of each year's cash flow must be evaluated separately by dividing it by the factor

$$\frac{E^c(t)}{E^c(0)}$$

for that year.

Year	Cash Flow ($)	$\dfrac{E^c(0)}{E^c(t)}$	Present Value ($)
0	−35,000	1.000	−35,000
1	5,500	0.966	5,314
2	5,500	0.916	5,037
3	5,500	0.868	4,774
4	5,500	0.855	4,705
5	5,500	0.842	4,634
6	5,500	0.810	4,457
7	5,500	0.779	4,283
8	5,500	0.749	4,120
	Total:	6.786	2,324

The investment earns the necessary return and appears to be worthwhile. The $2334 net present value is small compared to the initial investment, and thus the investment is fairly marginal.

If desired, the problem can be solved in dollar terms. It is necessary to first inflate the cash flows with the factor

$$\frac{\bar{p}(t)}{\bar{p}(0)}$$

and then find the present value by using the factor

$$\frac{E^d(t)}{E^d(0)}$$

Year	Cash Flow ($)	$\bar{p}(t)/\bar{p}(0)$	Cash Flow ($)	$\dfrac{E^d(0)}{E^d(t)}$	Present Value ($)
0	−35,000	1.000	−35,000	1.000	−35,000
1	5,500	1.120	6,160	0.863	5,314
2	5,500	1.210	6,655	0.757	5,037
3	5,500	1.282	7,051	0.677	4,774
4	5,500	1.372	7,546	0.623	4,705
5	5,500	1.482	8,151	0.589	4,634
6	5,500	1.615	8,883	0.502	4,457
7	5,500	1.760	9,680	0.442	4,283
8	5,500	1.919	10,555	0.390	4,120
			Total:	4.823	2,324

The present value results are identical to those calculated before, which is as expected since the reference time for the constant dollar is at time zero.

Either of these results could also be expressed as an annual cash flow if desired. The dollar equivalent uniform annual cash flow can be found by summing the values of the terms

$$\frac{E^d(0)}{E^d(t)}$$

as has already been done in the preceding tabular list. However, the values of $E(O)/E(t)$ were included in the summations and must be subtracted before determining the equivalent annual cash flows to be consistent with Equation 11.13. The equivalent dollar cash flow is then

$$ACF^d = \frac{\$2324}{3.823} = \$608$$

The equivalent constant dollar annual cash flow is

$$ACF^c = \frac{\$2324}{5.786} = \$402$$

Although there is nothing wrong with a detailed projection of inflation and time value as is used in this example, it should be pointed out that it does little to improve the quality of the analysis here. There is considerable uncertainty in the future price of the scrap metal. Much more uncertainty in the final economics is introduced from this source than any error introduced by using just an average rate of inflation and average time values.

Example 11.3

A particular company establishes their cost of money by averaging the cost from different sources for the previous four years. A time value of money of 13.5% has been established by applying this method, and it is requested that this value be used in economic analysis. The inflation rate varied from 7 to 12% during this four year period. The inflation rate is presently 12%. The company projects the inflation rate to drop two percentage points each year for the next three years and then stay constant at about 6%. How should economic calculations be carried out using this information?

These two statements, a constant apparent time value of 13.5% and an inflation rate declining from 12 to 6%, are contradictory. A change of this magnitude in the inflation rate would almost certainly result in a substantial change in the apparent time value. Furthermore, it is doubtful that the 13.5% cost of money is an accurate reflection of the cost of money. The inflation rate changed substantially over the four year period on which it is based. The dollar cost of money probably changed significantly during this time. Since the present inflation rate is at the high end of the range of inflation rates during the period, the 13.5% probably underestimates the present cost of money. These discrepancies should be resolved if at all possible before an economic analysis is made.

APPENDIX 11.1   NONCONSTANT INFLATION RATES AND EQUIVALENCY

The purpose of this appendix is to show that the only way one can use both constant real time value and constant apparent time value and still ensure consistent equivalence calculations is to also have a constant inflation rate as is described by Equation 3.15. This result should be no surprise since Equation 3.15 was derived to maintain consistency. However, when it was derived it was shown only that this relationship was a sufficient condition for consistency in general; it was not shown that it is also a necessary condition. This later derivation follows.

The following derivation uses the present value form of the equivalence calculation. It applies equally well to other expressions of equivalence. Consider a set of dollar cash flows Y_j^d that are arbitrary and occur one time period apart. The present value of these cash flows using a constant apparent time value is

$$PV^d = \sum_{j=0}^{n} Y_j^d \cdot \left[\frac{1.0}{1 + i_a} \right]^j \qquad \text{(A11.1-1)}$$

where i_a is the apparent time value per time period that separates the cash flows. Now consider the corresponding set of cash flows Y_j^c that is the constant dollar representation of the cash flows Y_j^d. The present value of these cash flows, using a constant real time value, is

$$PV^c = \sum_{j=0}^{n} Y_j^c \cdot \left[\frac{1.0}{1 + i_r} \right]^j \qquad \text{(A11.1-2)}$$

The relationship between the dollar cash flows and the constant dollar cash flows is

$$Y_j^c = Y_j^d \frac{\bar{p}(t_0)}{\bar{p}(t_j)} \qquad \text{(A11.1-3)}$$

where

$$t_j = \Delta t \cdot j$$

and Δt is the time between cash flows. This definition for t_j means that the reference time t_0 occurs at the time of the present value. Another time could be used, but this choice will simplify the calculations and will not affect the validity of the derivation. Since the reference time for the constant dollars and the time for the present values are the same, then

$$PV^d = PV^c \qquad \text{(A11.1-4)}$$

Equations A11.1-1 and A11.1-2 can now be equated and Equation A11.1-3 used to substitute for Y_j^c:

$$\sum_{j=0}^{n} Y_j^d \cdot \left[\frac{1.0}{1 + i_a} \right]^j = \sum_{j=0}^{n} Y_j^d \cdot \frac{\bar{p}(t_0)}{\bar{p}(t_j)} \cdot \left[\frac{1.0}{1 + i_r} \right]^j \qquad \text{(A11.1-5)}$$

The equation may be rearranged to

$$\sum_{j=0}^{n} Y_j^d \cdot \left\{ \frac{\bar{p}(t_0)}{p(t_j)} \cdot \left[\frac{1.0}{1 + i_r} \right]^j - \left[\frac{1.0}{1 + i_a} \right]^j \right\} = 0 \quad \text{(A11.1-6)}$$

Since Y_j^d is arbitrary, the only way to ensure that Equation A11.1-6 holds true in general is for

$$\frac{\bar{p}(t_0)}{\bar{p}(t_j)} \cdot \left[\frac{1.0}{1 + i_r} \right]^j - \left[\frac{1.0}{1 + i_a} \right]^j = 0 \qquad \text{(A11.1-7)}$$

for

$$0 \le j \le n$$

Equation A11.1-7 can be rearranged to

$$\left[\frac{1 + i_a}{1 + i_r} \right]^j = \frac{\bar{p}(t_j)}{\bar{p}(t_0)} \qquad \text{(A11.1-8)}$$

Since i_a and i_r are both to be constants, then

$$\left[\frac{1 + i_a}{1 + i_r} \right] = K \qquad \text{(A11.1-9)}$$

where K is a constant, and it can be said that

$$\bar{p}(t_j) = \bar{p}(t_0)K^j \qquad \text{(A11.1-10)}$$

The relationship between the price levels for any two time periods is

$$\frac{\bar{p}(t_j)}{\bar{p}(t_{j-1})} = \frac{\bar{p}(t_0)K^j}{\bar{p}(t_0)K^{j-1}} = K \qquad \text{(A11.1-11)}$$

Recall that the definition of the inflation for time period j is

$$f_j \equiv \frac{\bar{p}(t_j) - \bar{p}(t_{j-1})}{\bar{p}(t_{j-1})} = \frac{\bar{p}(t_j)}{\bar{p}(t_{j-1})} - 1$$

Substituting into the definition for inflation with Equation A11.1-11 gives

$$f_j = K - 1 \qquad \text{(A11.1-12)}$$

and since K is constant for the period

$$0 \le j \le n$$

then f_j must be constant for the same period and it may simply be replaced with

$$f_j = f \qquad\qquad (A11.1-13)$$

This result is sufficient to meet the original objective of this derivation. However, it may be carried a couple steps further. Substituting Equation A11.1-13 into A11.1-12 gives

$$f = K - 1 \qquad\qquad (A11.1-14)$$

which may be substituted into Equation A11.1-9 to give

$$1 + f = \frac{1 + i_a}{1 + i_r} \qquad\qquad (A11.1-15)$$

thus confirming Equation 3.15.

The conclusion of this derivation is that it is necessary to have a constant inflation rate over the period of an analysis if consistent equivalency calculations are to be guaranteed and the real and apparent time values are held constant. This result should surprise no one.

APPENDIX 11.2   NONCONSTANT INFLATION RATES AND ALTERNATIVE COMPARISON

It was shown in Appendix 11.1 that nonconstant inflation rates could not be used and still guarantee consistency between constant dollar-real time value and dollar-apparent time value equivalency calculations when the apparent and real time values are both held constant. Engineering economics deals with the investment of capital and the comparison of alternatives. This appendix carries the preceding derivation further to show that any of the three evaluation methods—present value, annual cash flow, and rate of return—may yield inconsistent decisions between the constant dollar-real time value and dollar-apparent time value approaches again with varying inflation but constant real and apparent time value.

A11.2-1 Present Value

To show that inconsistent decisions can be made, it is necessary to show that for a single set of cash flows one approach will yield a positive present value and the other approach, a negative present value. The present values by the two methods as described in Appendix 11.1 are

$$PV^d = \sum_{j=0}^{n} Y_j^d \cdot \left[\frac{1.0}{1 + i_a} \right]^j \qquad \text{(A11.2-1)}$$

and

$$PV^c = \sum_{j=0}^{n} Y_j^d \cdot \frac{\bar{p}(t_0)}{\bar{p}(t_j)} \cdot \left[\frac{1.0}{1 + i_r} \right]^j \qquad \text{(A11.2-2)}$$

The difference between the two present values is

$$PV^d - PV^c = \sum_{j=0}^{n} Y_j^d \cdot \left\{ \left[\frac{1.0}{1 + i_a} \right]^j - \frac{\bar{p}(t_0)}{\bar{p}(t_j)} \cdot \left[\frac{1.0}{1 + i_r} \right]^j \right\} \qquad \text{(A11.2-3)}$$

Since Y_j^d are arbitrary, they can be varied as desired for this derivation. The value of Y_0^d may be varied to any value and not affect the difference since

$$\left[\frac{1.0}{1 + i_a} \right]^0 - \frac{\bar{p}(t_0)}{\bar{p}(t_0)} \cdot \left[\frac{1.0}{1 + i_r} \right]^0 = 0 \qquad \text{(A11.2-4)}$$

for any reasonable values of i_a and i_r. If the inflation rate is not constant

$$\left[\frac{1.0}{1 + i_a} \right]^j - \frac{\bar{p}(t_0)}{\bar{p}(t_j)} \cdot \left[\frac{1.0}{1 + i_r} \right]^j$$

must be nonzero for at least one value of j for the difference to be nonzero. Let this value of j be designated as k, and let

$$\left[\frac{1.0}{1 + i_a} \right]^k - \frac{\bar{p}(t_0)}{\bar{p}(t_j)} \cdot \left[\frac{1.0}{1 + i_r} \right]^k = L \qquad \text{(A11.2-5)}$$

Now the other values of Y_j^d may be selected so that

$$PV^d - PV^c = 0$$

when

$$Y_k^d = 0$$

Now, Y_k^d may be varied arbitrarily so that the difference is nonzero and

$$PV^d - PV^c = Y_k^d \cdot L$$

The value of Y_0^d does not affect the difference and since it is arbitrary, it may be selected so that

$$PV^d = x$$

where x can be any value. The present value in constant dollars is then

$$PV^c = x - Y_k^d \cdot L$$

Since Y_k^d can be varied to any value and then x independently varied to any value, it is possible to let PV^d and PV^c take on any value independently. It is then possible to select values of Y_k^d and Y_0^d such that

$$PV^d < 0$$

while at the same time

$$PV^c > 0$$

and vice versa.

The conclusion to be drawn from this derivation is that it is entirely possible that alternatives and investments evaluated by present values may be shown to be desirable by one method and not by the other. It is not possible to guarantee the same conclusion by present value evaluation by using the constant dollar-real time value and dollar-apparent time value methods when the inflation rate varies and the apparent time value and the real time value are held constant.

A11.2-2 Annual Cash Flow

The annual cash flow can be related directly to the present value for each method:

$$ACF^d = \left(\frac{A}{P}, i_a, n \right) \cdot PV^d \qquad \text{(A11.2-7)}$$

$$ACF^c = \left(\frac{A}{P}, i_r, n \right) \cdot PV^c \qquad \text{(A11.2-8)}$$

The values of $(A/P, i, n)$ will differ in the two cases; however, both will be positive as long as

$$i > -1$$

and n is finite. It was just shown that PV^d could be positive while PV^c is negative. In this case, ACF^d will be positive while ACF^c is negative. The opposite situation also holds true; PV^d can be negative while PV^c is positive, and ACF^d will be negative and ACF^c positive. Either way, it is clear that it is possible to have a single set of cash flows that can give contradictory results between ACF^d and ACF^c when inflation varies and the real time value and the apparent time value are held constant.

A11.2-3 Rate of Return

We limit the discussion for the rate of return evaluation to sets of cash flows that are initially negative and then become positive without reversing again. It must first be argued that this limitation does not invalidate the conclusions drawn in the discussion of present value evaluation. Start with such a conventional set of cash flows for which the present value is zero and with a constant inflation rate. Both PV^d and PV^c will be zero in this case. The inflation is then described by

$$\frac{\bar{p}(t_j)}{\bar{p}(t_0)} = [1 + f]^j \qquad \text{(A11.2-9)}$$

and the summation in Equation A11.2-3 will be zero for every term. Now let inflation deviate from this relationship such that

$$\frac{\bar{p}(t_k)}{\bar{p}(t_0)} < [1 + f]^k \qquad \text{(A11.2-10)}$$

for this one term only. Also let k be sufficiently large that it refers to a positive cash flows. Only the kth term in the summation in Equation A11.2-3 now contributes to the difference:

$$PV^d - PV^c$$

The difference may be made to take on any positive value by selecting the appropriate positive value for Y_k^d. All other terms in the summation will not contribute to the difference and may be varied as desired to make PV^d take on any value as long as the sign of a cash flow is not changed. Since there are both positive and negative cash flows, PV^d may be increased or decreased. The difference may be made negative by reversing Inequality A11.2-10. Thus the limitation placed on the type of cash flows does not affect the results derived earlier. It is possible to arbitrarily vary the values of PV^d and PV^c.

It was shown in Appendix 7.1 that for the type of cash flows used here, the present value depends on the time value according to

$$\frac{\partial PV}{\partial i} < 0 \qquad \text{(A11.2-11)}$$

The rate of return for a set of cash flows is found by

$$0 = PV(i) = \sum_{j=0}^{n} Y_j \cdot \left[\frac{1.0}{1+i}\right]^j \qquad \text{(A11.2-12)}$$

where either dollar or constant dollar terms may be used. The resulting interest rate will be the apparent or real rate of return, respectively. The criterion for determining the desirability of a set of cash flows was

$$R_a > i_a$$

or

$$R_r > i_r$$

and is described in Chapter 7. Because of Inequality A11.2-11,

$$PV^d > 0 \Rightarrow r_a > i_a$$

and

$$PV^c < 0 \Rightarrow R_r < i_r$$

It has already been shown that a single cash flow set can exist such that the inequalities on the left side of these statements are true. It is then possible that an apparent rate of return evaluation may show a set of cash flows to be worthwhile whereas a real rate of return evaluation will show them not to be worthwhile. The converse can also be shown to be possible. Therefore, constant dollar-real rate of return and dollar-apparent rate of return evaluations may yield inconsistent results when inflation is not held constant.

This discussion can be taken one step further. It is not proper to compare mutually exclusive alternatives on the basis of their respective rates of return, but rather the rate of return of the incremental cash flow must be evaluated. However, many people like to compare the rates of return on independent alternatives to determine which ones to invest in. Consider three sets of cash flows. The first set has cash flows such that

$$PV_1^d = PV_1^c = 0$$

at the respective rates of return R_{a_1} and R_{r_1}. It has been shown that a second set of cash flows may exist such that

$$PV_2^d > PV_1^d \qquad\qquad \text{(A11.2-13)}$$

and

$$PV_2^c < PV_1^c \qquad\qquad \text{(A11.2-14)}$$

Inequality A11.2-13 implies that

$$R_{a2} > R_{a1} \qquad\qquad \text{(A11.2-15)}$$

since $PV_1^d = 0$. Inequality A11.2-14 implies that

$$R_{r2} < R_{r1} \qquad\qquad \text{(A11.2-16)}$$

It has also been shown that a third set of cash flows may exist such that

$$PV_3^d < PV_1^d \qquad\qquad \text{(A11.2-17)}$$

and

$$PV_3^c > PV_1^c \qquad\qquad \text{(A11.2-18)}$$

It then follows that

$$R_{a3} < R_{a1} \qquad\qquad \text{(A11.2-19)}$$

and

$$R_{r3} > R_{r1} \qquad\qquad \text{(A11.2-20)}$$

Comparing the rates of return for the three sets of cash flows, it is seen that

$$R_{a2} > R_{a3} \qquad\qquad \text{(A11.2-21)}$$

while

$$R_{r2} < R_{r3} \qquad\qquad (A11.2\text{-}22)$$

The dollar-apparent rate of return evaluation shows that alternative 2 has a higher rate of return than alternative 3, whereas the constant dollar-real rate of return evaluation shows just the converse. Thus inconsistent evaluations of relative rates of return are also possible when the inflation rate is not constant.

CHAPTER TWELVE

Inflation Relationships in Continuous Form

Just as all cash flows do not necessarily occur at the end of discrete time periods, price changes do not necessarily occur at the end of those same time periods. Prices may change at any time, and different prices change at different times. As a consequence, the overall price level $\bar{p}(t)$ changes more or less continuously throughout a year or other time period. The year end convention results in some calculation error for those cash flows that do not occur at the year end. The error is due to the time value from the time between occurrence of the cash flow and the year end and also possibly due to price changes during this same time. The problems encountered with discrete, one-year time periods were discussed in Chapter 4.

Discrete representation of time value and inflation is traditional but not necessary. Expression of these parameters in continuous form is equally valid. Continuous forms of the parameters eliminate the year end problem and also simplify the calculations for many evaluations. Cash flows at any point in time can be converted to equivalent cash flows at any other point in time. Conversions between dollars and constant dollars may also be made at any point in time. High inflation rates can aggravate cash flow timing errors, and continuous expression of both time value and inflation may be necessary in this case. Even when the continous representation is not needed for accuracy, it is often desirable to use it for convenience.

12.1 CONTINUOUS TIME VALUE

There are several ways to derive an expression for continuous interest. The simplest is to just hypothesize a time value relationship as follows:

$$Y(t_1) = Y(t_2) \cdot e^{i' \cdot [t_1 - t_2]} \qquad (12.1)$$

where $Y(t_1)$ and $Y(t_2)$ are equivalent cash flows at times t_1 and t_2, respectively, and i' is the time value in continuous form with units of inverse time. Comparing Equations 12.1 to the earlier expression of time value in Equation 3.11, it is seen that for Equation 12.1 to be valid, then

$$e^{i' \cdot [t_1 - t_2]} = [1.0 + i]^{[(t_1 - t_2)/\Delta t]}$$

(12.2)

where i is the corresponding time value for the discrete time period Δt. Taking the natural logarithm of both sides of this equation gives

$$i' = \frac{1}{\Delta t} \cdot \ln(1.0 + i)$$

(12.3)

Exponentiating both sides yields

$$i = e^{i' \cdot \Delta t} - 1.0$$

(12.4)

Equations 12.3 and 12.4 show that i and i' are unique functions of each other, so Equation 12.1 is a mathematically valid form of expressing equivalence and Equations 12.3 and 12.4 show the relationship between discrete and continuous time value.

The six traditional interest factors in discrete form are presented in Appendix C. These may be readily converted to continuous form. The term

$$[1.0 + i]^n$$

may be replaced with

$$e^{i' \cdot \Delta t \cdot n}$$

in these expressions, and the term i may be replaced with

$$e^{i' \cdot \Delta t} - 1.0$$

The resulting statement of the interest factors are presented in Table 12.1. These traditional interest factors are somewhat cumbersome to use with continuous time value. It is also necessary to specify Δt for these interest factors where Δt is the time period between cash flows. In some respects this requirement is an advantage since the cash flow frequency can be specified independently of the interest rate.

It is also possible to incorporate continuous cash flows into interest factors. A continuous cash flow is one that provides cash continuously at some rate, for example, \$2500 per week. Many cash flows that occur very frequently are best represented in this manner. The present value of such a continuous cash flow is found by

$$PV(t) = \int_{t_1}^{t_2} e^{-i' \cdot [\tau - t]} \cdot C(\tau) \cdot d\tau \qquad (12.5)$$

where C is the cash flow rate, the present value is calculated for the cash flow from t_1 to t_2 and τ is an integrating parameter. If the present value is taken at the beginning of the cash flow, then

$$t = t_1$$

and if the cash flow rate is constant, Equation 12.3 reduces to

$$PV = C \cdot \int_{t_1}^{t_2} e^{-i' \cdot [\tau - t_1]} \cdot d\tau \qquad (12.6)$$

When integrated, this equation yields

$$PV = C \cdot \frac{1 - e^{-i' \cdot t}}{i'} \qquad (12.7)$$

where t is defined here as

$$t = t_2 - t_1$$

The future value of this continuous cash flow can be found at time t_2 by applying the first interest factor in Table 12.1 to the present value in Equation 12.7. The result is

$$FV = C \cdot \frac{e^{i' \cdot t} - 1}{i'} \qquad (12.8)$$

Equations 12.7 and 12.8 may be used to generate four additional interest factors for use with continuous cash flows. Also, the first two interest factors in Table 12.1 may be expressed in more convenient terms. These interest factors are given in Table 12.2. The timing relationship for the cash flows in these interest factors are shown in Figure 12.1.

Continuous cash flows which do not stay constant can also be evaluated if

TABLE 12.1 Traditional Interest Factors in Continuous Form

$$\left(\frac{F}{P}, i', n, \Delta t\right) = e^{i' \cdot \Delta t \cdot n}$$

$$\left(\frac{P}{F}, i', n, \Delta t\right) = e^{-i' \cdot \Delta t \cdot n}$$

$$\left(\frac{F}{A}, i', n, \Delta t\right) = \frac{e^{i' \cdot \Delta t \cdot n} - 1.0}{e^{i' \cdot \Delta t} - 1.0}$$

$$\left(\frac{A}{F}, i', n, \Delta t\right) = \frac{e^{i' \cdot \Delta t} - 1.0}{e^{i' \cdot \Delta t \cdot n} - 1.0}$$

$$\left(\frac{P}{A}, i', n, \Delta t\right) = \frac{e^{i' \cdot \Delta t \cdot n} - 1.0}{[e^{i' \cdot \Delta t} - 1.0] \cdot e^{i' \cdot \Delta t \cdot n}}$$

$$\left(\frac{A}{P}, i', n, \Delta t\right) = \frac{[e^{i' \cdot \Delta t} - 1.0] \cdot e^{i' \cdot \Delta t \cdot n}}{e^{i' \cdot \Delta t \cdot n} - 1.0}$$

TABLE 12.2 Interest Factors for Continuous Cash Flows

$$\left(\frac{F}{C}, i', t\right) = \frac{e^{i' \cdot t} - 1.0}{i'}$$

$$\left(\frac{C}{F}, i', t\right) = \frac{i'}{e^{i' \cdot t} - 1.0}$$

$$\left(\frac{P}{C}, i', t\right) = \frac{1 - e^{-i' \cdot t}}{i'}$$

$$\left(\frac{C}{P}, i', t\right) = \frac{i'}{1 - e^{-i' \cdot t}}$$

$$\left(\frac{F}{P}, i', t\right) = e^{i' \cdot t}$$

$$\left(\frac{P}{F}, i', t\right) = e^{-i' \cdot t}$$

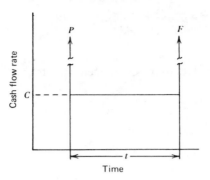

Figure 12.1 Cash flow timing for interest factors in Table 12.2.

necessary. Equation 12.5 can be used whenever $C(t)$ can be described. However, the evaluation can become mathematically complicated. Most problems can be adequately described without this added complexity.

12.2 CONTINUOUS INFLATION

A continuous inflation rate may be derived in exactly the same manner as for time value. The basic inflation relationship in continuous form is

$$\frac{\bar{p}(t)}{\bar{p}(t_0)} = e^{f' \cdot [t - t_0]} \qquad (12.9)$$

where f' is the continuous inflation rate and has units of inverse time. The relationship between the continuous and the discrete inflation rate is the same as for time value:

$$f' = \frac{1}{\Delta t} \cdot \ln(1 + f) \qquad (12.10)$$

and

$$f = e^{f' \cdot \Delta t} - 1.0 \qquad (12.11)$$

The time period Δt will normally be one year, but it can be most anything.

The relationship between real and apparent time value and the inflation rate given in Equation 3.15 is repeated:

$$1 + i_r = \frac{1 + i_a}{1 + f}$$

Each of these parameters may be replaced with its continuous counterpart to yield

$$e^{i'_r} = \frac{e^{i'_a}}{e^{f'}}$$

which reduces to

$$i'_a = i'_r + f' \qquad\qquad (12.13)$$

where the subscripts a and r refer to apparent and real time values, respectively, just as with the discrete forms.

Once the time values and the inflation rate are established in continuous form, engineering economics calculations are made pretty much as they are with discrete forms. The only difference is the flexibility that is possible in cash flow timing. All the other rules and limitations still apply. Present value comparisons are made exactly as before. Annual cash flow comparisons may be made in terms of equivalent uniform annual cash flows or in terms of equivalent uniform continuous cash flows, in either dollars or constant dollars, if desired. The rate of return in continuous form is again calculated by setting the present value to zero. Since

$$\frac{\partial i}{\partial i'} = e^{i' \cdot \Delta t} \qquad\qquad (12.14)$$

then

$$\frac{\partial i}{\partial i'} > 0 \qquad\qquad (12.15)$$

It then follows that for where

$$\frac{\partial PV}{\partial i} < 0 \qquad\qquad (12.16)$$

that

$$\frac{\partial PV}{\partial i'} < 0 \qquad\qquad (12.17)$$

and thus when

$$R' > i' \qquad (12.18)$$

where R' is the rate of return in continuous form, then

$$PV > 0 \qquad (12.19)$$

The conclusion then is that the rate of return in continuous form may be used to evaluate investments just as the discrete form is used.

The discussion thus far in this chapter pertains only to the situation of constant parameters (time value and inflation rate). The more general situation of varying parameters discussed in Chapter 11 can also utilize the continuous expressions of time value and the inflation rate. Interest factors are no longer usable in this case. The functions

$$\frac{\bar{p}(t)}{\bar{p}(t_0)}$$

and

$$\frac{E^d(t)}{E^d(t_0)}$$

or

$$\frac{E^c(t)}{E^c(t_0)}$$

are just as easily expressed in continuous form as in discrete form, and cash flow timing is not restricted to discrete time steps.

12.3 EXAMPLES

Example 12.1

A firm that manufactures a variety of heat exchangers is negotiating a contract with another company to supply a particular design of heat exchanger for a product they sell. The company has offered to make a contract to buy 1200 heat exchangers each year for three years at a fixed price of $4750 per unit. Furthermore, the company will agree to pay for an entire calendar year's purchases on January 1 of that year.

Engineers for the heat exchanger manufacturer have completed a cost analysis on this particular design. A special fabrication machine will need to be installed to manufacture the heat exchanger. The machine will cost $3,750,000 installed at today's prices and is available on relatively short notice. A special metal alloy tube must be used in the heat exchanger. Checking with suppliers of these tubes shows that they will cost $1240 per heat exchanger. The suppliers will not guarantee future prices, but past history shows that products fabricated from this metal have increased in price at about the same rate as overall inflation. The preferred supplier of these tubes is not a local manufacturer, and it will be necessary to buy the tubes in batches. It is anticipated that tubes for three months' production will be purchased at the beginning of each quarter. Other materials for the heat exchanger will cost about $300 per unit. They consist of a number of different items and can be expected to increase in cost on the average at about the same rate as inflation. Operating costs of the machinery, other expenses, and indirect overhead costs (but not including direct labor) will amount to about $670 per unit at today's prices and also should increase at about the same rate as overall inflation. There will be an estimated 50 direct hours of labor required for each exchanger. Plant labor presently earns $11.60/h. The workers have a contract with the firm that includes a cost of living adjustment. The wages are raised on July 1 each year to be consistent with the cost of living at that point in time. The wages continue at this level until the next July 1. It is presently near the first of the year.

Company economists predict the inflation rate to average 12% annually for the next three years. The company presently requires a 17% annual before tax, apparent rate of return on new product expansion investments. On the basis of this information, does this look like a good contract?

This looks like a complicated problem but is certainly no more difficult than one would expect for a real problem. The first thing to do is to summarize the cash flows:

1. Initial cost of new equipment: $3,750,000 at time zero in today's dollar.

2. Sales income: $5,700,000 in dollars at the beginning of each year.

3. Alloy tubes: $372,000 in today's dollar at beginning of each quarter.

4. Other materials: $360,000 in today's dollar paid throughout year.

5. Other expenses: $804,000 in today's dollar paid throughout year.

6. Direct labor: $580 per unit in dollars with midyear wage increases to the current price index at that time.

There are several ways to assess this investment. The easiest way is to determine the net real present value of each cash flow and add the results together to determine whether the venture has a positive net present value. The 17% required apparent rate of return must be converted to both real and continuous forms. The equivalent continuously compounded required apparent rate of return is

$$i'_a = \frac{1.0}{1.0 \text{ yr}} \cdot \ln(1.0 + 0.17) = 0.157/\text{yr}$$

The inflation rate in continuous form is

$$f' = \frac{1.0}{1.0 \text{ yr}} \cdot \ln(1.0 + 0.12) = 0.113/\text{yr}$$

The real required rate of return in continuous form is

$$i'_r = 0.157/\text{yr} - 0.113/\text{yr} \equiv 0.044/\text{yr}$$

Now each set of cash flows is examined in detail to determine its net present value:

1. The present value of the new equipment is simply $-\$3,750,000$.
2. The sales income is a uniform series of dollar payments starting at time zero. The net present value can be determined by using a conventional interest factor and the apparent required rate of return in discrete form: $PV = \$5,700,000 + (P/A, 17\%, 2) \cdot \$5,700,000 = \$14,736,000$
3. The alloy tube expense cash flow is a little more difficult. The cost of tubes for the first quater is $372,000 at time zero. This expense will remain constant in terms of today's dollars. The net present value can be calculated by using the appropriate interest factor with a continuous real rate of return:
$PV = -\$372,000 - (P/A, 0.044/\text{yr}, 11, \tfrac{1}{4}\text{yr}) \cdot \$372,000$
$PV = -\$372,000 - 10.304 \cdot \$372,000$
$PV = -\$4,205,000$
4. The other materials can be considered a continuous cash flow in today's dollars at the rate of $360,000/yr. The net present value can be found by using the real continuous required rate of return and the

appropriate continuous interest factor:

$$P = -\$360,000/\text{yr} \cdot (P/C, 0.044/\text{yr}, 3 \text{ yr}) = -\$1,012,000$$

5. The other expenses are handled in a similar manner:

$$P = -\$804,000 \cdot (P/C, 0.044/\text{yr}, 3 \text{ yr}) = -\$2,261,000$$

6. The labor cash flow is probably the most difficult. The most direct approach, but not necessarily the most elegant, is to calculate the wage rate for the appropriate periods and then create a series of cash flows. Assuming that the cost of living and inflation are the same for our purposes, the following wage rates will apply:

1st half, 1st year—$11.60/h
2nd half, 2nd year—$11.60/h $\cdot$ $e^{0.113/\text{yr} \cdot 0.5 \text{ yr}}$ = $12.28/h
1st half, 2nd year—$12.28/h
2nd half, 2nd year—$11.60/h $\cdot$ $e^{0.113/\text{yr} \cdot 1.5 \text{ yr}}$ = $13.75/h
1st half, 3rd year—$13.75/h
2nd half, 3rd year—$11.60/h $\cdot$ $e^{0.113/\text{yr} \cdot 2.5 \text{ yr}}$ = $15.40/h

Fifty hours of labor are required for each unit, and 1200 units are to be produced per year. These give a working rate of 60,000 h/yr. The continuous cash flow rate for each period can now be established:

1st half, 1st year $C = \$11.60/\text{h} \cdot 60,000 \text{ h/yr} = \$696,000/\text{yr}$
2nd half, 1st year $C = \$12.28/\text{h} \cdot 60,000 \text{ h/yr} = \$737,000/\text{yr}$
1st half, 2nd year $C = \$12.28/\text{h} \cdot 60,000 \text{ h/yr} = \$737,000/\text{yr}$
2nd half, 2nd year $C = \$13.75/\text{h} \cdot 60,000 \text{ h/yr} = \$825,000/\text{yr}$
1st half, 3rd year $C = \$13.75/\text{h} \cdot 60,000 \text{ h/yr} = \$825,000/\text{yr}$
2nd half, 3rd year $C = \$15.40/\text{h} \cdot 60,000 \text{ h/yr} = \$924,000/\text{yr}$

These cash flows are in dollars. All that is necessary now is to use the appropriate interest factors with the apparent time value in continuous form:

$$\begin{aligned}
PV = &-(P/C, 0.157/\text{yr}, 0.5 \text{ yr}) \cdot \$696,000/\text{yr} - (P/F, 0.157/\text{yr}, 0.5 \text{ yr}) \cdot \\
&(P/C, 0.157/\text{yr}, 1.0 \text{ yr}) \cdot \$737,000/\text{yr} - (P/F, 0.157/\text{yr}, 1.5 \text{ yr}) \cdot \\
&(P/C, 0.157/\text{yr}, 1.0 \text{ yr}) \cdot \$825,000/\text{yr} - (P/F, 0.157/\text{yr}, 2.5 \text{ yr}) \cdot \\
&(P/C, 0.157/\text{yr}, 0.5 \text{ yr}) \cdot \$924,000/\text{yr} = -\$1,869,000
\end{aligned}$$

The total net present value can now be calculated. Since the constant dollar expressions use the present as a reference time, the constant dollar and dollar present values may be added:

$$\begin{aligned}
P_{\text{total}} = &\ \$14,736,000 - \$3,750,000 - \overline{\$}4,205,000 - \overline{\$}1,012,000 \\
&-\overline{\$}2,261,000 - \$1,869,000 \\
= &\ \$1,639,000
\end{aligned}$$

The venture does have a positive net present worth. This result indicates that the venture should be profitable. Note, however, that a venture of this magnitude probably would justify additional economic analysis. It may be desirable to calculate the actual rate of return for the project. Such a calculation would be best done with the aid of a computer. Also, a thorough sensitivity and risk analysis is likely to show that the profitability of this venture is strongly dependent on the future inflation rate since income is fixed in dollars but that expenses will increase as prices increase.

This example was not intended to show that calculations with required rates of return and inflation in continuous form are tedious and complicated. Its purpose is to show that proper analysis of very complicated sets of cash flows are easier when the continuous forms are used. Solution of this problem this accurately with discrete forms would have been difficult. In general, use of the continuous forms will not make the problem more difficult. Rather, they usually will make it simpler and will result in few approximations being made. It may be argued that the errors caused by using discrete time periods and discrete interest and inflation relationships are usually small compared to other uncertainties in most engineering economic analysis. This statement may be true, but the ease with which continuous relationships may be used tends to indicate that the main reason why they are not more widely used is tradition.

APPENDIX A

Deflation

The text has dealt only with the problem of making valid engineering economic analyses when inflation is prevalent. The opposite problem of making economic analyses during periods of deflation has not been addressed. However, a review of the theory in the text shows that it is not applicable only to situations of inflation but also to the more general situation of the varying value of the dollar. There are no assumptions made anywhere in the development of the theory that says that the value of the dollar must decrease and that the prices must increase. The material thus applies to deflation as well as inflation. The orientation is toward inflation because it is the presently existing situation and has been the predominantly occurring situation in recent history and because there is no real evidence that it will not be the situation for sometime in the future. However, it may be worthwhile to briefly discuss how the theory may be applied to economic analysis during deflationary times.

Deflation may be regarded simply as negative inflation. The inflation rate was previously defined as

$$ f = \frac{\bar{p}(t_1) - \bar{p}(t_0)}{\bar{p}(t_0)} \tag{A.1} $$

where f is the inflation rate for the time period

$$ \Delta t = t_1 - t_0 $$

With deflation, the overall price levels will decline and the inflation rate will be negative. It is possible to apply the theory presented in the text to deflation without further modification simply by expressing the deflation as negative inflation. A more convenient way to express deflation is in the form of a deflation rate that is analogous to the inflation rate. This deflation rate is defined by

$$d = \frac{\bar{p}(t_0) - \bar{p}(t_1)}{\bar{p}(t_0)} \tag{A.2}$$

where d is the deflation rate. The deflation rate is the fractional (or percentage) decrease in the overall price level over a time period Δt just as the inflation rate is the fractional increase in the overall price level.

Comparing Equations A.1 and A.2, it is evident that

$$d = -f \tag{A.3}$$

The negative of the deflation rate may be substituted for the inflation rate into any of the equations derived previously. Generally, the inflation rate shows up in the equations in the term

$$[1 + f]$$

and this may simply be replaced with the term

$$[1 - d]$$

when dealing with deflation. This term is present in many equations. Three of the more basic relationships are Equation 2.6 for the price levels, which becomes

$$\frac{\bar{p}(t_0)}{\bar{p}(t)} = \frac{1}{[1 - d]^n} \tag{A.4}$$

where

$$n = \frac{t - t_0}{\Delta t}$$

Equation 3.15 for the relationship between real and apparent time value, which becomes

$$[1 + i_a] = [1 + i_r] \cdot [1 - d] \tag{A.5}$$

and Equation 4.3 for projecting individual prices, which becomes

$$\frac{p_j^d(t)}{p_j^d(t_0)} = [1 - d]^n \cdot p_j^r \left(\frac{t}{t_0}\right) \tag{A.6}$$

where again

$$n = \frac{t - t_0}{\Delta t}$$

Most other relationships result from these three equations and the principle of time value either in dollar or constant dollar terms. Note that the real time value is greater than the apparent time value with deflation.

Many people associate inflation and deflation with economic prosperity and economic recession, respectively. Some economic theories indicate that this relationship is true; others disagree. There have been instances in the past where the relationship appears to have been true, and there have been instances when it has not. The theory presented in this text includes no assumptions about any such relationship. It deals with how to account for changing value of the dollar and not with what causes that change. To be sure, the economic strategies called for during periods of economic prosperity differ from those required during periods of economic recession. Such considerations are more appropriately included in setting objectives of the firm and in major capital allocation decisions. These considerations may ultimately affect the time value for a firm and consequently the type of calculations covered in this text. The determination of objectives and capital allocation are complicated topics and are outside the scope of the material in this text. The problem of variations in the value of the dollar thus appears to be independent of questions of strategy for prosperity and recession. However, these two subjects become interrelated when one considers the latter question in greater detail.

APPENDIX B

Nomenclature

Characters

A	magnitude of a uniform series of cash flows
ACF	annual cash flow
C	continuous cash flow
d	depreciation for a single item; deflation rate
D	deflator, a factor for converting from dollars to constant dollars; depreciation; demand curve (Appendixes 2.2 and 3.1)
e	base of the natural logarithm
E	expenses; equivalent value function
f	inflation rate
F	future amount by equivalence
G	gross income
i	time value; interest rate
I	price index; interest
IC	initial cost
$\ln$	natural logarithm
L	intermediate variable for derivation in Appendix 11.2
n	number of time periods; economic life
O	rate of output of goods and services for an economy
p	price; real price (Appendixes 2.2 and 3.1)
$\bar{p}$	average price level
P	present amount by equivalence
PC	present cost of an alternative
PV	present value of a set of cash flows

Q	quantity of goods and services; value of some goods and services (Appendix 3.1)
r	tax rate
R	return; internal rate of return
s	fractional annual increase in a series of cash flows
S	principle of an investment; supply curve (Appendixes 2.2 and 3.1)
SV	salvage value
SYD	sum of years digits
t	time
T	tax
TI	taxable income
V	value placed on a quantity of wealth; quantity of value
x	arbitrary quantity
X	quantity of wealth
Y	cash flow
τ	integrating parameter (time)

Symbols

$\$$	dollars
$\bar{\$}$	constant dollars

Subscripts

a	apparent
at	after tax
b	base time (price index)
e	effective
I	income
0	reference time (constant dollar)
s	stated (interest rate)
t	time value; true time value (Appendix 3.1)

Superscripts

c	constant dollars
d	dollars

r   relative

'   used to denote continuous forms of f, i, and R; also used to denote
             a set of goods and services for a price index; it is used several
             other places simply to denote a different set of conditions or
             different representation.

APPENDIX C

Interest Factors

The widespread availability of digital computers and powerful hand-held calculators is rapidly making lengthy interest tables unnecessary for engineering economics calculations. It is often easier to calculate the interest factors from the mathematical equations than to look up a value in a table. This is particularly true when interest rates do not match up nicely with the tables, as is the rule rather than the exception in actual practice. Even if an interest rate is used that conforms to a table, it will often need to be changed to a nonconforming interest rate when converted from real to apparent or vice versa. The mathematical relationships for the six common interest factors are presented in Table C.1. For those who prefer to use tabulated values, complete tables may be found in many texts on engineering economics.

TABLE C.1   Mathematical Expressions for Common Interest Factors

Compound amount factor

$$\left(\frac{F}{P}, i, n\right) = [1 + i]^n$$

Present worth factor

$$\left(\frac{P}{F}, i, n\right) = \frac{1}{[1 + i]^n}$$

Series compound amount factor

$$\left(\frac{F}{A}, i, n\right) = \frac{[1 + i]^n - 1}{i}$$

Series present worth factor

$$\left(\frac{P}{A}, i, n\right) = \frac{[1 + i]^n - 1}{i \cdot [1 + i]^n}$$

Sinking fund factor

$$\left(\frac{A}{F}, i, n\right) = \frac{i}{[1 + i]^n - 1}$$

TABLE C.1 *(continued)*

Capital recovery factor

$$\left(\frac{A}{P}, i, n\right) = \frac{i \cdot [1 + i]^n}{[1 + i]^n - 1}$$

APPENDIX D

Price Indexes

TABLE D.1 Gross National Product Implicit Price Deflator

Year	GNP-IPD[a]		Year	GNP-IPD
1947	62.9		1964	92.0
1948	67.2		1965	94.1
1949	66.6		1966	97.2
1950	67.8		1967	100.0
1951	72.5		1968	104.6
1952	73.4		1969	109.7
1953	74.6		1970	115.7
1954	75.6		1971	121.5
1955	77.2		1972	126.6
1956	79.6		1973	133.9
1957	82.3		1974	146.8
1958	83.7		1975	161.0
1959	85.4		1976	169.2
1960	87.0		1977	179.4
1961	87.7		1978	192.4
1962	89.4		1979	209.6
1963	90.6		1980	224.6

[a]Average for year; base year is 1967.
Source: *Business Conditions Digest*, U.S. Department of Commerce, Bureau of Economic Analysis.

TABLE D.2 Consumer Price Index

Year	CPI[a]	Year	CPI
1947	66.9	1964	92.9
1948	72.1	1965	94.5
1949	71.4	1966	97.2
1950	72.1	1967	100.0
1951	77.8	1968	104.0
1952	79.5	1969	109.8
1953	80.1	1970	116.3
1954	80.5	1971	121.3
1955	80.2	1972	125.3
1956	81.4	1973	133.1
1957	84.3	1974	144.7
1958	86.6	1975	161.2
1959	87.7	1976	170.5
1960	88.7	1977	181.5
1961	89.6	1978	195.4
1962	90.6	1979	217.4
1963	91.7	1980	246.8

[a]Average for year; base year 1967.

Source: *Business Condition Digest*, U.S. Department of Commerce, Bureau of Economic Analysis.

TABLE D.3 Consumer Price Index[a]

Date		All Items	Food and Beverages	Housing	Fuels and Utilities	Apparel and Upkeep	Transportation	Medical Care	Entertainment[b]	Other
1967	June	99	99	99	100	100	100	99	99	99
	December	101	100	100	100	102	102	102	101	103
1968	June	104	103	103	101	105	103	105	104	105
	December	106	105	105	103	109	103	109	106	107
1969	June	110	109	109	104	111	108	113	108	109
	December	112	112	113	105	115	108	115	110	114
1970	June	116	115	119	107	116	113	121	113	116
	December	119	115	123	111	119	117	124	116	119
1971	June	122	119	124	115	120	120	129	119	120
	December	123	120	127	118	122	119	130	121	123
1972	June	125	123	129	120	122	120	132	123	126
	December	127	126	131	122	124	121	135	124	126
1973	June	132	140	134	127	127	125	137	126	129
	December	139	151	141	136	131	127	141	128	131
1974	June	147	160	149	149	136	141	149	134	136
	December	155	170	160	158	142	144	159	140	144
1975	June	161	174	166	167	141	150	168	144	147
	December	166	181	172	176	143	158	177	148	150
1976	June	170	181	177	182	147	166	184	151	153
	December	174	182	182	192	152	171	192	154	156

TABLE D.3 (continued)

Date		All Items	Food and Beverages	Housing	Fuels and Utilities	Apparel and Upkeep	Transportation	Medical Care	Entertainment[b]	Other
1977	June	182	194	189	202	154	179	202	158	158
	December	186	192	192	208	158	179	209	171	178
1978	June	195	209	202	218	160	186	218	176	181
	December	203	214	212	220	163	193	228	181	189
1979	June	217	229	225	239	166	213	238	188	194
	December	230	236	244	255	172	228	251	193	204
1980	June	248	246	267	282	177	250	265	205	213
	December	258	259	277	290	184	261	276	212	225

[a]Based on average value for 1967.
[b]Recreation and reading prior to 1977.
Source: U.S. Department of Labor, Bureau of Labor Statistics.

TABLE D.4 Wholesale Price Index—All Items

Year	WPI[a]	Year	WPI
1947	76.5	1964	94.7
1948	82.8	1965	96.6
1949	78.7	1966	99.8
1950	81.8	1967	100.0
1951	91.1	1968	102.5
1952	88.6	1969	106.5
1953	87.4	1970	110.4
1954	87.6	1971	113.9
1955	87.8	1972	119.1
1956	90.7	1973	134.7
1957	93.3	1974	160.1
1958	94.6	1975	174.9
1959	94.8	1976	183.0
1960	94.9	1977	194.2
1961	94.5	1978	209.3
1962	94.8	1979	235.4
1963	94.5	1980	268.8

[a]Average for year, based year 1967, referred to as Producer Price Index beginning in 1978.
Source: *Business Conditions Digest*, U.S. Department of Commerce, Bureau of Economic Analysis.

TABLE D.5 Wholesale Price Index Commodity Classifications[a]

Code	Item
01	Farm products
011	Fresh and dried fruits and vegetables
012	Grains
013	Livestock
014	Live poultry
015	Plant and animal fibers
016	Fluid milk
017	Eggs
018	Hay, hayseeds, and oilseeds
019	Other farm products
02	Processed foods and feeds

TABLE D.5 *(continued)*

Code	Item
021	Cereal and bakery products
022	Meats, poultry, and fish
023	Dairy products
024	Processed fruits and vegetables
025	Sugar and confectionery
026	Beverages and beverage materials
0271	Animal fats and oils
0272	Crude vegetable oils
0273	Refined vegetable oils
0274	Vegetable oil end products
028	Miscellaneous processed foods
029	Manufactured animal feeds
03	Textile products and apparel
031	Cotton products (Synthetic fibers after 1975)
032	Wool products (Processed yarns and threads after 1975)
033	Manmade fiber textile products (Gray fabrics after 1975)
034	Finished fabrics
035	Apparel (see 0381)
036	Textile house furnishings (see 0382)
037	Miscellaneous textile products
0381	Apparel (beginning in 1978)
0382	Textile house furnishings (beginning in 1978)
04	Hides, skins, leather, and related products
041	Hides and skins
042	Leather
043	Footwear
044	Other leather and related products
05	Fuels and related products and power
051	Coal
052	Coke
053	Gas fuels
054	Electric power
056	Crude petroleum
057	Petroleum products, refined

Code	Item
06	Chemicals and allied products
061	Industrial chemicals
0621	Prepared paint
0622	Paint materials
063	Drugs and pharmaceuticals
064	Fats and oils, inedible
065	Agricultural chemicals and chemical products
066	Plastic resins and materials
067	Other chemicals and allied products
07	Rubber and plastic products
071	Rubber and rubber products
0711	Crude rubber
0712	Tires and tubes
0713	Miscellaneous rubber products
0721	Plastic construction products
0722	Unsupported plastic film and sheeting
0723	Laminated plastic sheets, high pressure
08	Lumber and wood products
081	Lumber
082	Millwork
083	Plywood
084	Other wood products
09	Pulp, paper, and allied products
091	Pulp, paper, and allied products excluding building paper, and board
0911	Woodpulp
0912	Wastepaper
0913	Paper
0914	Paperboard
0915	Converted paper and paperboard products
092	Building paper and board
10	Metals and metal products
101	Iron and steel
102	Nonferrous metals
103	Metal containers
104	Hardware

TABLE D.5 *(continued)*

Code	Item
105	Plumbing fixtures and brass fittings
106	Heating equipment
107	Fabricated structural metal products
108	Miscellaneous metal products
11	Machinery and equipment
111	Agricultural machinery and equipment
112	Construction machinery and equipment
113	Metalworking machinery and equipment
114	General purpose machinery and equipment
116	Special industry machinery and equipment
117	Electrical machinery and equipment
119	Miscellaneous machinery
12	Furniture and household durables
121	Household furniture
122	Commercial furniture
123	Floor coverings
124	Household appliances
125	Home electronic equipment
126	Other household durable goods
13	Nonmetallic mineral products
1311	Flat glass
132	Concrete ingredients
133	Concrete products
134	Structural clay products excluding refractories
135	Refractories
136	Asphalt roofing
137	Gypsum products
138	Glass containers
139	Other nonmetallic minerals
14	Transportation equipment
141	Motor vehicles and equipment
144	Railroad equipment
15	Miscellaneous products

Code	Item
151	Toys, sporting goods, small arms, ammunition
152	Tobacco products
153	Notions
154	Photographic equipment and supplies
159	Other miscellaneous products

[a]Producer Price Index Beginning 1978.

TABLE D.6 Wholesale Price Indexes[a,b]

Code[c]	1970	1971	1972	1973	1974	1975	1976	1977	1978	1979 June	1979 Dec.	1980 June	1980 Dec.
01	111	113	125	176	188	187	191	193	213	243	243	233	265
011	112	120	128	168	192	184	178	192	218	226	211	224	245
012	99	101	103	184	258	224	206	165	183	219	228	215	265
013	117	118	143	190	171	188	173	173	220	264	253	240	251
014	100	100	104	180	157	190	167	175	200	183	195	167	219
015	90	93	118	198	194	153	224	202	193	220	222	247	294
016	115	119	122	145	173	180	201	203	220	244	264	266	291
017	127	101	104	166	161	160	179	162	159	171	198	147	218
018	99	109	118	220	229	200	210	234	216	258	230	207	310
019	117	115	125	147	164	170	223	326	275	281	319	309	296
02	112	114	121	148	171	183	178	186	203	221	229	234	251
021	108	111	115	134	171	178	172	173	190	206	224	233	249
022	116	116	130	166	164	191	182	182	217	241	243	227	248
023	111	115	119	131	146	156	169	173	188	208	220	230	243
024	110	114	120	130	155	170	170	187	203	221	222	227	237
025	116	119	122	132	259	254	191	177	198	213	234	325	335
026	113	116	118	122	141	162	174	201	200	208	222	234	238
0271	140	131	127	230	328	342	210	267	291	320	291	257	296
0272	121	129	108	175	291	208	163	198	219	250	227	180	205
0273	118	135	115	154	266	213	188	199	229	231	194	153	217
0274	112	121	121	144	225	212	174	199	209	220	231	229	237

	1	2	3	4	5	6	7	8	9	10	11	12	13
028	113	113	115	123	159	178	175	190	199	211	220	223	241
029	104	104	116	199	184	172	194	205	197	220	225	205	247
03	107	109	114	124	139	138	148	154	160	168	173	182	190
031	106	111	122	144	175	—	102^d	107^d	110^d	119^d	125^d	135^d	142^d
032	99	94	99	128	119	—	100^d	101^d	102^d	109^d	113^d	122^d	128^d
033	102	111	108	122	136	—	105^d	105^d	119^d	125^d	133^d	134^d	143^d
034	—	—	—	—	—	—	101^d	104^d	104^d	107^d	109^d	116^d	120^d
035	111	113	115	119	130	133	140	147	—	—	—	—	—
036	104	104	109	113	143	152	159	172	—	—	—	—	—
037	107	117	127	125	171	—	—	—	—	—	—	—	—
0381	—	—	—	—	—	—	—	—	152	160	162	172	177
0382	—	—	—	—	—	—	—	—	179	189	197	203	219
04	110	114	131	143	145	149	168	179	200	267	249	241	257
041	104	115	214	254	196	175	258	287	361	611	444	316	393
042	108	113	140	160	154	152	188	201	239	415	325	284	332
043	113	117	125	131	140	148	159	169	183	220	227	232	237
044	106	108	118	130	137	141	153	163	177	212	208	216	224
05	106	114	119	134	208	245	266	302	323	393	489	575	612
051	150	182	194	218	332	386	369	389	430	452	458	467	476
052	127	149	156	167	248	331	347	379	412	431	431	431	431
053	103	108	114	127	162	217	287	388	429	519	671	750	842
054	105	114	122	129	163	193	208	233	251	270	287	321	338
056	106	113	114	126	211	246	254	274	300	356	471	549	596
057	101	107	109	129	223	258	277	308	321	423	555	681	716
06	102	104	104	110	147	181	187	193	199	219	238	262	268

TABLE D.6 *(continued)*

Code[c]	1970	1971	1972	1973	1974	1975	1976	1977	1978	1979 June	1979 Dec.	1980 June	1980 Dec.
061	101	102	101	103	152	207	219	224	226	259	292	327	335
0621	112	116	118	119	146	167	174	182	192	201	211	237	242
0622	101	102	104	113	152	177	190	206	212	237	255	274	281
063	101	102	103	104	113	127	134	141	148	159	164	173	182
064	133	134	116	228	338	255	250	279	316	374	327	256	316
065	88	92	92	97	138	204	188	188	198	209	233	258	263
066	91	89	89	92	144	181	194	198	200	230	263	288	274
067	109	112	114	118	148	169	171	176	182	191	202	226	234
07	109	109	109	112	136	150	159	168	175	193	206	217	224
071	110	112	114	118	138	152	163	174	185	204	224	238	246
0711	102	99	99	112	139	146	160	172	187	220	239	263	268
0712	109	109	109	111	133	149	162	170	179	198	223	235	243
0713	114	118	121	125	141	156	164	177	190	203	217	230	237
0721	97[e]	95[e]	93[e]	94[e]	119[e]	124[e]	127[e]	133[e]	136[e]	147[e]	148[e]	155[e]	154[e]
0722	—	101[f]	99[f]	100[f]	132[f]	149[f]	155[f]	160[f]	163[f]	175[f]	185[f]	192[f]	194[f]
0723	—	99[f]	98[f]	98[f]	116[f]	125[f]	131[f]	141[f]	147[f]	160[f]	165[f]	173[f]	178[f]
08	114	127	144	177	184	177	206	236	276	300	290	280	299
081	114	136	159	205	207	193	233	277	322	355	339	313	333
082	116	121	128	144	157	160	177	194	235	259	250	253	273
083	109	115	131	155	161	161	187	212	236	238	238	242	264
084	117	119	125	150	167	162	166	184	212	239	241	239	236
09	108	110	113	122	152	170	179	187	196	217	231	251	257

091	109	110	114	123	153	172	181	187	196	218	233	253	259
0911	109	112	112	128	218	283	286	282	267	309	340	388	393
0912	125	112	134	197	266	110	185	187	191	207	221	207	191
0913	111	114	116	121	149	173	182	194	206	228	243	258	270
0914	101	102	106	115	152	170	176	177	179	200	215	243	241
0915	108	110	114	122	145	162	170	177	186	207	220	239	245
092	101	103	106	113	124	127	139	157	187	181	184	209	219
10	117	119	124	133	172	186	196	209	227	258	274	282	291
101	115	122	128	136	179	201	216	230	254	283	293	303	316
102	125	116	117	135	187	172	182	195	208	257	291	291	294
103	113	122	129	135	165	192	202	218	243	268	281	303	303
104	111	117	120	125	141	163	173	185	200	217	227	240	250
105	113	116	120	126	149	162	174	187	199	217	226	249	254
106	111	116	118	120	135	151	158	166	174	186	195	205	213
107	112	118	122	127	161	189	194	207	227	249	258	270	279
108	114	119	124	130	157	181	187	196	212	231	240	251	258
11	111	116	118	122	139	161	171	182	196	212	223	239	250
111	113	117	122	126	144	169	183	198	213	228	243	256	270
112	116	121	126	131	152	185	190	214	233	253	268	287	301
113	114	117	120	126	147	172	183	199	217	239	255	275	286
114	114	119	122	127	151	179	190	202	217	234	246	264	275
116	116	121	124	130	151	175	188	203	223	246	256	256	291
117	106	110	110	112	125	141	147	154	165	177	187	187	209
119	113	117	120	124	140	162	172	181	195	207	216	228	239
12	108	110	111	115	128	140	146	152	160	169	177	185	192

TABLE D.6 *(continued)*

Code[e]	1970	1971	1972	1973	1974	1975	1976	1977	1978	1979 June	1979 Dec.	1980 June	1980 Dec.
121	112	115	117	123	137	146	154	162	173	185	194	202	210
122	114	118	120	129	152	167	174	186	202	222	225	236	242
123	100	99	99	102	115	125	131	136	142	147	153	162	170
124	105	107	108	109	118	132	139	145	153	160	165	175	178
125	94	94	93	92	93	94	91	88	89	90	88	89	91
126	116	121	126	130	149	169	179	190	203	220	252	266	285
13	113	122	126	130	153	174	186	201	223	247	259	283	291
1311	116	124	122	121	129	139	150	161	173	183	186	194	203
132	115	122	127	131	149	172	187	199	217	242	250	272	279
133	112	121	126	132	152	171	180	192	214	244	253	276	278
134	110	114	117	123	135	151	164	180	197	217	227	230	234
135	121	127	129	136	144	166	184	200	217	234	249	267	274
136	103	126	131	136	196	226	238	253	292	324	343	401	395
137	100	107	115	121	138	144	154	184	229	251	255	257	253
138	120	132	135	139	156	180	195	214	245	266	274	295	325
139	112	124	127	128	189	220	233	251	276	303	342	395	416
14	105[g]	110[g]	114[g]	115[g]	126[g]	142[g]	151[g]	161[g]	173[g]	187[g]	195[g]	202[g]	224[g]
141	109	115	118	119	129	145	154	164	176	190	198	204	226
144	115	121	129	135	164	201	217	234	253	273	289	306	324

15	110	113	115	120	133	148	154	164	185	204	227	257	265
151	109	113	114	118	132	146	150	155	163	175	184	197	207
152	114	117	118	122	133	150	163	180	199	214	226	245	254
153	108	112	112	114	137	151	162	172	182	190	197	217	225
154	105	106	107	108	128	131	136	140	146	152	165	203	207
159	108	112	116	125	142	156	153	167	213	254	308	359	372

[a]Producer Price Indexes Beginning 1978.
[b]All indexes are referenced to 100 for the year of 1967 unless otherwise noted. Indexes are annual averages for the years 1970–1978.
[c]See Table D.5 for explanation of codes.
[d]Index referenced to 100 for December 1975.
[e]Index referenced to 100 for December 1969.
[f]Index referenced to 100 for December 1970.
[g]Index referenced to 100 for Decmeber 1968.

Source: U.S. Department of Labor, Bureau of Labor Statistics.

Index